Wearable electronics and photonics

Related titles from Woodhead's textile technology list:

Smart fibres, fabrics and clothing (ISBN 1 85573 546 6)

This important book provides a guide to the fundamentals and latest developments in smart technology for textiles and clothing. The contributors represent a distinguished international panel of experts and the book covers many aspects of cutting edge research and development. It examines the background to smart technology and goes on to cover a wide range of the material and fibre science aspects of the technology.

Handbook of technical textiles (ISBN 1 85573 385 4)

This major handbook looks at the manufacture, processing and applications of hi-tech textiles for a huge range of applications including: heat and flame protection; waterproof and breathable fabrics; textiles in filtration; geotextiles; medical textiles; textiles in transport engineering and textiles for extreme environments. It is an essential guide for textile yarn and fibre manufacturers; producers of woven, knitted and non-woven fabrics; textile finishers; designers and specifiers of textiles for new or novel applications as well as lecturers and graduate students on university textile courses.

Details of these books and a complete list of Woodhead's textile technology titles can be obtained by:

- visiting our web site at www.woodheadpublishing.com
- contacting Customer Services (e-mail: sales@woodhead-publishing.com; fax: +44(0) 1223 893694; tel: +44(0) 1223 891358 ext. 30; address: Woodhead Publishing Ltd, Abington Hall, Abington, Cambridge CB1 6AH, England.

Wearable electronics and photonics

Edited by
Xiaoming Tao

The Textile Institute

CRC Press
Boca Raton Boston New York Washington, DC

WOODHEAD PUBLISHING LIMITED
Cambridge England

Published by Woodhead Publishing Limited in association with The Textile Institute
Abington Hall, Abington
Cambridge CB1 6AH, England
www.woodheadpublishing.com

Published in North America by CRC Press LLC
2000 Corporate Blvd, NW
Boca Raton FL 33431, USA

First published 2005, Woodhead Publishing Ltd and CRC Press LLC
© 2005, Woodhead Publishing Ltd
The authors have asserted their moral rights.

Every effort has been made to trace and acknowledge ownership of copyright. The publishers will be glad to hear from the copyright holders whom it has not been possible to contact concerning Table 7.1.

This book contains information obtained from authentic and highly regarded sources. Reprinted material is quoted with permission, and sources are indicated. Reasonable efforts have been made to publish reliable data and information, but the authors and the publishers cannot assume responsibility for the validity of all materials. Neither the authors nor the publishers, nor anyone else associated with this publication, shall be liable for any loss, damage or liability directly or indirectly caused or alleged to be caused by this book.

Neither this book nor any part may be reproduced or transmitted in any form or by any means, electronic or mechanical, including photocopying, microfilming and recording, or by any information storage or retrieval system, without permission in writing from the publishers.

The consent of Woodhead Publishing and CRC Press does not extend to copying for general distribution, for promotion, for creating new works, or for resale. Specific permission must be obtained in writing from Woodhead Publishing or CRC Press for such copying.

Trademark notice: Product or corporate names may be trademarks or registered trademarks, and are used only for identification and explanation, without intent to infringe.

British Library Cataloguing in Publication Data
A catalogue record for this book is available from the British Library.

Library of Congress Cataloging in Publication Data
A catalog record for this book is available from the Library of Congress.

The publishers' policy is to use permanent paper from mills that operate a sustainable forestry policy, and which has been manufactured from pulp which is processed using acid-free and elementary chlorine-free practices. Furthermore, the publishers ensure that the text paper and cover board used have met acceptable environmental accreditation standards.

Woodhead Publishing ISBN 1 85573 605 5
CRC Press ISBN 0-8493-2595-1
CRC Press order number: WP2595

Typeset by Ann Buchan (Typesetters), Shepperton, Middlesex
Printed by TJ International, Padstow, Cornwall, England

Contents

	Contributor contact details	ix
	Preface	xiii
1	**Introduction**	**1**
	XIAOMING TAO, The Hong Kong Polytechnic University, Hong Kong	
1.1	Overview	1
1.2	Current and future wearable technology	2
1.3	Applications of wearable electronics and photonics	8
1.4	Implications of wearable technology	10
1.5	References	12
2	**Electrostatically generated nanofibres for wearable electronics**	**13**
	FRANK K. KO, AFAF EL-AUFY and HOA LAM, Drexel University, USA and ALAN G. MACDIARMID, University of Pennsylvania, USA	
2.1	Introduction	13
2.2	Electrospinning process	15
2.3	Electroactive nanofibres	21
2.4	Ultra-low dielectric constant of nanocomposite fibrous film	34
2.5	Conclusions	37
2.6	Acknowledgements	38
2.7	References	39
3	**Electroceramic fibres and composites for intelligent apparel applications**	**41**
	HELEN LAI-WA CHAN, KUN LI and CHUNG LOONG CHOY, The Hong Kong Polytechnic University, Hong Kong	
3.1	Introduction	41

3.2	Fabrication of samarium and manganese doped lead titanate fibres	42
3.3	Fabrication of ceramic fibre/epoxy 1-3 composites	45
3.4	Electromechanical properties of ceramic fibre/epoxy 1-3 composites	49
3.5	The modified parallel and series model of ceramic/polymer 1-3 composites	49
3.6	Possible uses of ceramic fibres and composites in intelligent apparel applications	54
3.7	Acknowledgements	55
3.8	References	55

4	Electroactive fabrics and wearable man–machine interfaces	59
	DANILO DE ROSSI, FEDERICO CARPI, FEDERICO LORUSSI, ENZO PASQUALE SCILINGO and ALESSANDRO TOGNETTI, University of Pisa, Italy and RITA PARADISO, Smartex s.r.l., Italy	
4.1	Introduction	59
4.2	Sensing fabrics	62
4.3	Actuating fabrics	67
4.4	Smart fabrics for health care	71
4.5	Smart fabrics for motion capture	71
4.6	Smart textiles as kinaesthetic interfaces	76
4.7	Conclusions	79
4.8	Acknowledgements	79
4.9	References	79

5	Electromechanical properties of conductive fibres, yarns and fabrics	81
	PU XUE, XIAOMING TAO, MEI-YI LEUNG and HUI ZHANG, The Hong Kong Polytechnic University, Hong Kong	
5.1	Introduction	81
5.2	Conductive textiles	82
5.3	Electromechanical properties of PPy-coated conductive fibres/yarns	84
5.4	Performance of electrically conductive fabrics	95
5.5	Applications	101
5.6	Conclusions	102

5.7	Acknowledgement	103
5.8	References	103

6	Integration of fibre optic sensors and sensing networks into textile structures	105
	MAHMOUD EL-SHERIF, Drexel University, USA	
6.1	Introduction	105
6.2	Smart textiles	107
6.3	Modelling and analysis	111
6.4	Manufacturing of smart textiles	115
6.5	Applications of smart textiles	124
6.6	Acknowledgements	133
6.7	References	133
6.8	Bibliography	134

7	Wearable photonics based on integrative polymeric photonic fibres	136
	XIAOMING TAO, The Hong Kong Polytechnic University, Hong Kong	
7.1	Introduction	136
7.2	Photonic band-gap materials	136
7.3	Fibre-harvesting ambient light-reflective displays	138
7.4	Opto-amplification in active disordered media and photonic band-gap structures	140
7.5	Electroluminescent fibres and fabrics	145
7.6	Textile-based flexible displays	151
7.7	Acknowledgements	151
7.8	References	152

8	Communication apparel and optical fibre fabric display	155
	VLADAN KONCAR, ENSAIT-GEMTEX Laboratory, France and EMMANUEL DEFLIN and ANDRÉ WEILL, France Telecom Recherche et Développement, France	
8.1	Introduction	155
8.2	Communication apparel	156
8.3	Optical fibre fabric display	163
8.4	Acknowledgements	174
8.5	References	174

9	Wearable computing systems – electronic textiles	177
	TÜNDE KIRSTEIN, DIDIER COTTET, JANUSZ GRZYB and GERHARD TRÖSTER, Swiss Federal Institute of Technology Zurich, Switzerland	
9.1	Introduction	177
9.2	Why is clothing an ideal place for intelligent systems?	178
9.3	Electronic textiles	179
9.4	Electrical characterisation of textile networks	184
9.5	Conclusions	194
9.6	Future challenges	195
9.7	References	196
10	Data transfer for smart clothing: requirements and potential technologies	198
	JAANA RANTANEN and MARKO HÄNNIKÄINEN, Tampere University of Technology, Finland	
10.1	Introduction	198
10.2	Smart clothing concept model	199
10.3	Data transfer in smart clothing	202
10.4	Implementations for communication	214
10.5	Summary	220
10.6	References	220
11	Interaction design in smart textiles clothing and applications	223
	SHARON BAURLEY, University of the Arts London, UK	
11.1	Introduction	223
11.2	Knowledge age: dematerialisation of information and communications technology and the rise of ubiquitous intelligence	224
11.3	New commercial imperatives	226
11.4	Design and development: multidisciplinary collaboration	228
11.5	A new language for textiles: combining the real and the virtual	229
11.6	Technology enablers	236
11.7	Future technology enablers	239
11.8	Conclusions	240
11.9	Acknowledgement	241
11.10	References	241
11.11	Sources of further information	242
	Index	244

Contributor contact details

(* = main point of contact)

Chapter 1

Professor Xiaoming Tao
Institute of Textiles and Clothing
The Hong Kong Polytechnic University
Hung Hom, Kowloon
Hong Kong

Tel: (852) 2766 6470
Fax: (852) 2954 2521
E-mail: tctaoxm@inet.polyu.edu.hk

Chapter 2

Professor Frank K. Ko, Afaf El-Aufy and Hoa Lam*
Fibrous Materials Laboratory
Department of Materials Science and Engineering
Drexel University
31st and Market Street
Philadelphia, PA 19104, USA

Tel: (215) 895-1640
Fax: (215) 895-6760
E-mail: fko@coe.drexel.edu
   sg85c7f4@drexel.edu
   (Hoa Lam)

Professor Alan G. MacDiarmid
Department of Chemistry
University of Pennsylvania
Philadelphia, PA, USA

Tel: (215) 898-8307
Fax: (215) 898-8378
E-mail: macdiarm@a.chem.upenn.edu

Chapter 3

Professor Helen Lai-wa Chan,* Kun Li and Professor Chung Loong Choy
Department of Applied Physics
The Hong Kong Polytechnic University
Yuk Choi Road, Hung Hom, Hong Kong

Tel: (852) 2766 5692
Fax: (852) 2766 1202
E-mail: apahlcha@polyu.edu.hk

Chapter 4

Professor Danilo De Rossi,* Dr Federico Carpi, Dr Federico Lorussi, Dr Enzo Pasquale Scilingo and Dr Alessandro Tognetti
Interdepartmental Research Centre 'E. Piaggio'
Faculty of Engineering
University of Pisa
Via Diotisalvi, 2
56100 Pisa, Italy

Fax: +39 (0)50 2217051
Tel: +39 (0)50 2217050
E-mail: d.derossi@ing.unipi.it

Dr Rita Paradiso
Smartex s.r.l.
Via Giuntini, 13
56023 Navacchio
Pisa, Italy

Fax: +39 (0)50 754351
Tel: +39 (0)50 754350
E-mail: rita@smartex.it

Chapter 5

Dr Pu Xue*
Institute of Textiles and Clothing
The Hong Kong Polytechnic University
Hung Hom, Kowloon
Hong Kong

Tel: (852) 2766 6518
Fax: (852) 2773 1432
E-mail: tc389@inet.polyu.edu.hk

Professor Xiaoming Tao
(See Chapter 1)

Dr M. Y. Leung
Institute of Textiles and Clothing
The Hong Kong Polytechnic University
Hung Hom, Kowloon
Hong Kong

Tel: (852) 2766 6487
Fax: (852) 2773 1432
E-mail: tclens@inet.polyu.edu.hk

Mr H. Zhang
Institute of Textiles and Clothing
The Hong Kong Polytechnic University
Hung Hom, Kowloon
Hong Kong

Fax: (852) 2773 1432
E-mail: tczhhui@inet.polyu.edu.hk

Chapter 6

Mahmoud El-Sherif
Professor of Materials and Electrical
  and Computer Engineering
Director, Fibre Optics and Photonics
Manufacturing Engineering Center
3141 Chestnut Street
Drexel University
Philadelphia, PA 19104
USA

On leave of absence as:
President & CEO
Photonics Laboratories, Inc.
3619 Market Street
Philadelphia, PA 19104
USA

Tel: +001 (215) 387-9970
Fax: +001 (215) 387-4520
E-mail: melsherif@photonicslabs.com
Web site:
  http://www.photonicslabs.com

Chapter 7

Professor Xiaoming Tao
(See Chapter 1)

Chapter 8

Professor Vladan Koncar*
GEMTEX Laboratory
ENSAIT, Ecole Nationale Supérieure
  des Arts et Industries Textiles
9, rue de l'Ermitage
BP 30329
F-59056 Roubaix
France

Fax: +33320248406
Tel: +33320258959
E-mail: vladan.koncar@ensait.fr

Emmanuel Deflin and André Weill
France Telecom Recherche et Développement
28, chemin du Vieux Chêne
F-38243 Meylan
France

Fax: +33476764450
Tel: +33476762416
E-mail: emmanuel.deflin@france
  telecom.com

Chapter 9

Dr Tünde Kirstein,* Dr Didier Cottet, Dr Janusz Grzyb and Professor Gerhard Tröster
Wearable Computing Lab
Swiss Federal Institute of Technology Zurich
ETH Zentrum, IfE
Gloriastrasse 35
CH-8092 Zurich
Switzerland

Tel: +41 1 6322741
Fax: +41 1 6321210
E-mail: kirstein@ife.ee.ethz.ch
troester@ife.ee.ethz.ch

Chapter 10

Jaana Rantanen
Tampere University of Technology
Institute of Electronics
Korkeakoulunkatu 3, FIN-33720 Tampere
Finland

Tel: +358 3 3115 3401
Fax: +358 3 3115 2620
E-mail: jaana.rantanen@tut.fi

Dr Marko Hännikäinen*
Tampere University of Technology
Institute of Digital and Computer Systems
Korkeakoulunkatu 1, FIN-33720 Tampere
Finland

Tel: +358 3 3115 3837
Fax: +358 3 3115 4575
E-mail: marko.hannikainen@tut.fi

Chapter 11

Dr Sharon Baurley
School of Fashion and Textile Design
Central Saint Martins College of Art and Design
University of the Arts London
Southampton Row
London WC1B 4AP
UK

Fax: +44 20 7514 7050
Tel: +44 20 7514 8525
E-mail: s.baurley@csm.arts.ac.uk

Preface

This book is made up of contributions from a panel of international experts in wearable electronics and photonics and covers many aspects of cutting edge research and development. It comprises eleven chapters. Chapter 1 provides background information on wearable electronics and photonics and a brief overview of existing and emerging technologies. It also explains the structure of the book. Chapters 2 to 5 discuss topics related to materials and devices. Chapter 2, contributed by Professor Frank Ko, Afaf El-Aufy and Hoa Lam of Drexel University and Professor Alan MacDiarmid of Pennsylvania University, deals with electrostatically generated nanofibres for wearable electronics. Professor Helen Lai-wa Chan, Kun Li and Professor Chung Loong Choy of the Hong Kong Polytechnic University provide a detailed review of electroceramic fibres and composites in Chapter 3. Professor Danilo De Rossi and his colleagues from Pisa University write about electroactive fabrics and wearable man–machine interfaces in Chapter 4. Chapter 5 summarises recent developments by the editor's group in the fundamental aspects of electrically conductive fabric structures and puts together a few theoretical treatments of the electromechanical properties of various fabric structures.

Chapters 6, 7 and 8 are devoted to topics related to wearable photonics. Professor Mahmoud El-Sherif of Drexel University writes about embedded fibre optic sensors and integrated smart textile structures in Chapter 6. In Chapter 7 the editor presents a review of various flexible photonic display technologies and their development. Professor Vladan Koncar from ENSAIT describes communication apparel and optical fibre fabric displays in Chapter 8.

Chapters 9 and 10 focus on integrated structures and system architectures. Chapter 9 was contributed by a research group from the Swiss Federal Institute of Technology in Zurich. Here Dr Tünde Kirstein and her colleagues discuss wearable computing systems. Jaana Rantanen and Dr Marko Hännikäinen from Tampere University of Technology in Finland provide an overview of the requirements and potential technologies for data transfer in wearable electronics clothing in Chapter 10.

Chapter 11, written by Dr Sharon Baurley of Central Saint Martins College of Art and Design, describes various issues that fashion designers face when involved in the design and creation of wearable electronics and photonics.

This book provides a window through which a part of the exciting, emerging technology can be seen. The possibilities offered by wearable technology are remarkable and widespread. Even as this book was being prepared, many new advances were achieved around the world. It is the hope of the editor and contributors that this book will help researchers and designers to make their dreams a reality.

The editor is grateful to the Hong Kong Research Grants Council and The Hong Kong Polytechnic University for their partial funding support. In particular, the editor wishes to thank Dr Pu Xue for her assistance in compiling this book.

Xiaoming Tao

1
Introduction

XIAOMING TAO
The Hong Kong Polytechnic University, Hong Kong

1.1   Overview

Revolutionary changes have been occurring at an unprecedented rate in many fields of science and technology. The invention of electronic chips, computers, the internet, wireless communication, the discovery and complete mapping of the human genome, rapid advancements in nanotechnology, and many other developments, have transformed the entire world and affected nearly every human being on this planet. Looking ahead, the future promises even more. The technology of the future will have new features such as terascale, nanoscale, complexity, cognition and holism. The new capability of terascale takes us three orders of magnitude beyond the present general purpose and generally accessible computing capabilities. In a very short time, we will be connecting millions of systems and billions of information appliances to the internet. Technologies will develop to an incredible speed of over one trillion (1×10^9) operations per second. The technology in nanoscales will take us three orders of magnitude below the size of most of today's human-made devices. It will allow us to arrange atoms and molecules inexpensively in most of the ways permitted by physical laws. It will let us make supercomputers that fit on the head of a fibre; impart sensing and actuating mechanisms in micrometre- or nano-structures; allow wireless communication between devices, our body and environments; and make fashionable, intelligent clothing with built-in electronic and photonic functions.

The classical definition of electronics and photonics is the science and technology related to the generation, transmission, modulation and detection of electrons and photons, respectively. A wearable is a device that has the above functions, is always attached to a person and is comfortable and easy to keep and use. In other words, it is apparel with unobtrusively built-in electronic and photonic functions.

Wearable electronic and photonics have evolved from continuous technological advancements. An example is the evolution of a timing device. The carriage clock of three hundred years ago became a pocket watch and then a wristwatch. Now, personal electronic and photonic devices have been built into items that can be worn as jewellery and accessories. A recent development is the 'wrist camera watch'. This wearable digital camera is able to capture grey-scale photos anytime and store up to 100 pictures, view them instantly, or upload them to any computer via infrared red transmission. Another case is 'the wrist-type MP3 player', which can store and play up to 66 minutes of songs in MP3 (MPEG audio layer 3 or motion picture experts group audio layer 3) format downloaded from a computer using a universal serial bus (USB) connection. One more example is 'on hand PC', containing a 16-bit 3.6 MHz central processing unit (CPU) and 30 different built-in applications such as an address book, calculator, sound player and even games.

The first commercial range of wearable electronics apparel ICD+ was released in 2000 by Industrial Clothing Design Plus. It was co-produced by Philips NV and Levis Strauss with the collaboration of the designer Massimo Osti. One particular jacket design consists of ear gear and a microphone integrated into the collar with a simple 'body area network' made up of wires integrated into the design of the jacket. The jacket is also integrated with a global service mobile (GSM) phone and an MP3 player, which are operated using unified remote control. A book by Koninklijke Philips NV (2000) illustrates many such design concepts.

The examples mentioned above use quite simple and conventional technology but represent a step towards the incorporation of electronics into wearable items. There are at least three levels of sophistication in wearable technology: block-based technology, which connects all available devices and adds them to clothing as detachables; embedded technology, which is integrated into clothing by micro-electronic packaging technology; fibre-based technology, which are all devices in the form of fibres or fabrics. These examples belong to the first group of block-based systems.

Four years have passed since the introduction of the first commercial product in 2000. There are not many successful commercial products currently in the market. One of the major reasons for this is that most currently available technologies and materials are simply not suitable for commercial products. Electronics companies, together with university research laboratories, have devoted a great deal of effort and funding to developing the technological foundations for wearable electronics and photonics. It is the intention of the author of this chapter to provide an overview of various existing and emerging technologies for wearable electronics and photonics.

1.2   Current and future wearable technology

A typical system architecture design of a wearable electronic/photonic product is shown in Fig. 1.1. It comprises at least several basic functions: interface, communication, data management, energy management and integrated circuits.

1.2.1   Interface technologies

Devices such as sensors are often used to obtain information, for instance,

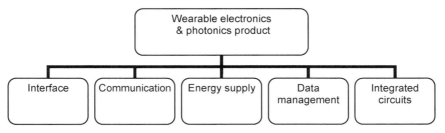

1.1 General system configuration of a typical wearable electronics and photonics product.

environmental sensors, physiological function sensors, antennae, global-positioning-systems receivers, cameras and sound sensors. The information needs to be processed somehow by the wearer. An interface is a suitable medium for transacting information between devices and the wearer as well as between the wearer and outside world.

Input interface

A wearer may input information to the devices, for example, to control which sensor to use. The most common input interface for this purpose involves buttons or keyboards because simple button interfaces are easy to learn, implement and use with few errors. As the complexity of wearable electronic devices increases, however, the need for more complex interfaces arises. Another input interface is voice recognition and writing pads. However, current technology has a number of drawbacks. First, the influence of background noise is so large that incorrect information may be processed. In addition current technology requires more processing power than previous technologies. Furthermore, current technology has difficulty recognising and distinguishing between different people's voices.

Fabric-based interface devices are very attractive for use in wearable electronics and photonics. Keyboards and buttons have been made from either multilayered woven circuits or polymer systems. Fabric-based sensors made from conductive fabrics or fibre optics have been used to measure movement and temperature. By designing appropriate fabric structures, various fabric electrodes and antennas have been successfully developed and applied to a few commercial products.

Output interface

Wearable devices have output interfaces by which information is presented to the wearer. Vibration (tactile) interfaces have been used. An example of this is the vibration function in mobile phones, by which the user is silently alerted to an incoming call. Many portable devices use audio interfaces. In both cases, the amount of information given is quite small. Voice synthesis (the opposite of voice recognition) via earphones is an alternative, as the wearer does not need to decode the message and can understand it directly. A third category of output interface is the visual interface. These include, for instance, seven-segment or dot matrix displays, liquid crystal displays (LCDs), organic and polymeric light-emitting diodes (OLEDs and PLEDs), and fibre optic displays (FODs). The displays may take two forms: wearable flat panel displays or head-mounted displays.

The main display technology used in portable electronics today is the LCD screen. It is neither flexible nor lightweight. Moreover, it can be bulky and its angle visibility is poor. Holographic polymer dispersed liquid crystals (HPDLCs) are still in their infancy; however, they may offer better performance in terms of flexibility. Polymer light-emitting diodes (PLEDs) are very promising candidates for future wearables, as they have high contrast, a high level of brightness, require much less power and are flexible. Flexible displays based on polymeric fibre optics are also being investigated by a number of researchers.

Electroactive polymer actuators take the form of fibres, yarns and structures based on thin film. They are used as artificial muscles for robotics. According to their actuating mechanisms, they can be broadly divided into two groups: electronic and ionic. The electronic polymers include electrostrictive, electrostatic, piezoelectric and ferroelectric polymers. They can hold induced displacement when a DC (direct current) voltage is applied and have a high level of energy density in air. However, a high activation field greater than 150 V μm^{-1} is required. Ionic polymeric materials include polymer metal composites, conducting polymers and polymer–carbon–nanotube composites. They normally perform actuation in a solution and have a low activation voltage of 1–5 V μm^{-1}. All of these actuators have limitations for use in wearable devices. A promising new technology is based on the dielectric elastomer, which is activated with low voltage in the air and is very robust and flexible. Books by Tao (2001) and Bar-Cohen (2001) provide very comprehensive accounts of dialectric elastomers.

1.2.2 Communication technologies

Communication refers to the transfer of information. This can be between two wearable devices on the user (short-range communications) or between two users via the internet or a network protocol (long-range communications).

Long-range communications

The technologies have been well developed in the mobile phone revolution. Portable devices such mobile phones or personal digital assistants (PDAs) have always used radio frequencies to enable communication. This is understandable, since it is just not viable to have long wires or optical links. A variety of communication systems are already available, the main one being GSM. While this system is presently very suitable for voice transmission, a substantial amount of research has recently been undertaken on allowing data to be transmitted (requiring a greater bit rate) through this system. GSM generally allows files or data to be transmitted and faxes to be sent at 9.6 kbps. Third-generation (3G) wireless systems are now being commercialised. They are capable of handling services of up to 384 kbps, sufficient to transfer pictures and videos. If the transmission of voice or low-content information is required, the present GSM system is sufficient.

Short-range communications

This is the area that particularly needs to be developed, since the techniques that are presently available are not adequate. Several approaches have been considered, including embedded wiring, infrared, Bluetooth technology and personal area network (PAN).

Embedded wiring is very cumbersome and constrictive to the user. Infrared, as used on remote controls, requires direct lines of sight to be effective but this would be difficult or impractical for devices located inside wallets, purses and pockets. Bluetooth technology is a newly developed technology. It is a new standard that will allow any sort of electronic equipment – for instance, from computers and cell phones to keyboards and headphones – to make its own connections without wires, cables, or any direct action from the user. Bluetooth components may interact without help from the user and are wireless. Another important advantage is their ability to minimise interference by sending very weak signals (limiting their range to about 10 m) and using a technique called spread spectrum hopping.

Personal area network or PAN was first developed by the MIT Media Lab in collaboration with IBM. This technology turns the human body into a network, taking advantage of the natural salinity of the human body, which makes it an excellent conductor of electrical current. PAN has a data transmission rate of 2400-bauds, sufficient to carry identification, financial or medical information but not good enough for audio or video information. A PAN rejects interference very well because it is mostly limited to transmission through the human body. However, touching a person equipped with a PAN is like tapping a phone line, i.e. security is a problem.

1.2.3 Data management technologies

The storing and processing of data are topics relating to the management of data. In wearable electronics and photonics, the storing of data is a problem that requires special attention. Storage technologies are used to keep information such as music, pictures or data banks. The following three storage technologies are the most commonly used. First are magnetic storage systems, from music tapes to hard disk drives, which are the most common form of storing information. This is due to their low cost and ease of use, and to their long MTBF (mean time before failure) of over

10 years. The second group consists of optical storage systems, which use a laser beam and optoelectronic sensors to read and store data. This technology has been the backbone of data storage for nearly two decades, with compact discs (CDs) (which are now rewritable using magneto-optic technology), and also digital versatile discs (DVDs) as the primary methods of storing data for music, software, personal computing and videos. The third are solid-state storage (flash memory storage) systems, the most recent medium of storage, which makes use of an EEPROM (electrically erasable programmable read-only memory) chip. Solid-state means that there are no moving parts – everything is electronic instead of mechanical. Their robustness, small size, weight and low power consumption make them well suited for application in wearable electronics.

Much research on more advanced means of storing information is based on magnetic and optic media. For instance, IBM manufactures a 1 GB (giga byte) magnetic drive as small as a compact flash card, but much cheaper. Optical storage using holographic memory can reach one terabyte (TB) of data in a sugar-cube-sized crystal, more than can be contained in 1000 CDs, while CDs and DVDs only make use of the surface area of the recording medium. Success in this kind of research would help spur growth in the wearable industry and the cost of producing wearable devices would fall.

1.2.4 Energy management technologies

The power supply for wearable electronics and photonics must be light and discreet, to be capable of being incorporated into clothing. Such a supply must be either long lasting or easy to recharge on the move. It should be robust enough to endure wearing and caring conditions. At present, batteries in the form of standard AA batteries or lithium batteries are the most common type of power source. Unless current levels of power consumption by electronic devices are reduced phenomenally or battery energy density increases drastically, other sources of power are likely to be required for wearable electronics and photonics.

A potential alternative lies in the ongoing miniaturisation of fuel cell technology. The smallest fuel cell so far has been developed by Toshiba and a team from MIT. It uses methanol, and is less than 2 inches (5 cm) long and weighs a fraction of an ounce. Similar to batteries, fuel cells generate electrical power by converting the chemical energy of a given type of fuel (e.g. hydrogen and oxygen) into electrical energy. However, fuel cells have far longer lives than conventional batteries of a similar size since oxygen does not need to be stored and only hydrogen is stored in metal hydrides or methanol. Recharging refills the fuel cells with hydrogen or methanol.

Another alternative is to harvest a small fraction of the kinetic energy from human movement. The kinetic Seiko watch has a miniature electric generator that is driven by arm movements, as well as a capacitor to store some charge so that the watch will run for intervals when the watch is stationary. Piezoelectric inserts in a shoe can harness walking power. Piezoelectric materials, such as PVDF (polyvinylidene fluoride), create an electrical charge when mechanically stressed. The deformation of the shoe during walking provides the necessary compression to generate power from piezoelectric piles. Apart from generating electricity, storage devices may be needed if the supply of kinetic power is used in electrical storage devices such as capacitors or in mechanical storage devices such as flywheels, pneumatic pumps or clock springs.

Other forms of power supply have been considered and investigated. Photovoltaic cells harvest the energy of the sun and semiconductor thermal couples generate electricity from the difference in temperature between the human body and the environment. These technologies have yet to produce sufficient power for wearable electronics. Alternatively, power may be transmitted to wearable devices remotely via microwaves, an area on which much research is being performed today. The greatest advantage of such an approach is that there would be a constant supply of power without the need for recharging.

1.2.5 Integrated circuits

Nowadays, most integrated circuits are made with silicon because of its superior semiconductor properties. Current chip fabrication processes exert limitations on the size of the chips or on the number of transistors on a chip. It has been speculated that a solution may lie in molecular electronics superseding silicon in the future. Molecular electronics employs devices based on a single molecule or single molecular wires to process signals and information. Molecules have the capability to conduct and transfer energy between one another and act like switches. If this process can somehow be manipulated and controlled, it would be possible to have these molecules and molecular structures perform tasks such as encoding, manipulating and storing information.

An important drawback is that silicon is not flexible. In contrast, conductive or semi-conductive polymeric materials are flexible, lightweight, strong and have a low production cost. These properties make them perfect for wearable electronics. The electronic properties of the conducting polymers may not match those of extremely pure and monocrystalline silicon. The polymeric chip would not be competing with the conventional silicon chip (at least not in the immediate future), but rather complementing it. At present, properties such as the switching speed and durability of the silicon chip are far superior to those of the polymeric chip, but the latter has the advantages of low price and flexibility (the ability to be folded double without affecting performance).

Wearable electronic devices might be incorporated into clothing, but still need to be connected together on the garment where the devices are allocated. A number of conductive fabrics have been made by using intrinsically conductive fibres or yarns. This allows the yarn to be sewn or embroidered with industrial machinery. Individual strands of yarns can be addressed so that a strip of this fabric can function like a ribbon cable. These fabrics have notable characteristics, including high conductivity, high tensile strength and good thermal stability.

1.3 Applications of wearable electronics and photonics

In the past few decades, many desk electronic appliances have been made portable because of constant miniaturisation in electronics. It is reasonable to assume that, in the future, some of these portable devices will become so small and convenient to carry that they will be wearable. Applications of the technology will be widespread and far reaching.

1.3.1 Information and communications

An evolution in lifestyles in recent years has led to increased mobility and, at the same time, a strong desire for instant access to information and communications. The tendency is towards ease of use/comfort: people want devices that are more unobtrusive and less inconvenient to use. A friendly and comfortable wearable computer or a wearable mobile phone integrated into some form of apparel will have a market if it is priced at an affordable level.

1.3.2 Health care and medical applications

In affluent societies, people are becoming increasingly health conscious. Meanwhile, the ageing of the population in many developed societies is bringing a heavy burden to bear on medical, especially hospital, systems as well as government budgets. Wearable electronics may provide personal systems for having our physiological status monitored. If necessary, medical advice can be given or treatment administrated anywhere, not just in a hospital, thereby leading to more mobility as well as to more efficient and effective health services.

Some noticeable examples can illustrate the applications. The Wearable Cardioverter Defibrillator by LifeCor Inc. has a chest harness and a hip pack, which provides immediate emergency medical aid to people prone to heart attacks. As soon as electrodes of the defibrillator sense the irregular beating of the heart, an audio warning is given before electricity is discharged. Then, the nearest hospital is notified. Another example is a wearable artificial kidney, which serves as a haemodialyser but has the advantage of being able to be fitted around the neck. In a form of undergarment, the wearable motherboard by Georgia Tech has sensors that are detachable so that they can be positioned at the right locations for users of different sizes. Such sensors can be used to monitor vital body signs (such as the heartbeat, respiration rate or temperature) of patients recovering from specific illnesses or to monitor patients at home rather than in a hospital. The Wearable Polysomnograph by Advanced Medical Corporation is a wearable ECG (electrocardiograph), which allows patients to be monitored at home with data being sent to a hospital via an internet connection.

Many people live their lives with a physical handicap, such as the loss of a sense like vision and hearing, or a loss of mobility in one part of their body or in all of it. Some future wearable electronic devices could help alleviate the suffering of such people. For example, wearable devices equipped with artificial muscles can be deployed to help limbs and arms become more mobile. Personal guidance systems are being developed to help the blind use the global positioning system (GPS) and the computer's geographical information system (GIS) to keep track of locations with the aid of a highly detailed map. Cochlear implants are being developed to help the deaf replace the lost functionality of damaged or missing hair cells by sending signals to the intact underlying nerve structure.

1.3.3 Fashion, leisure and home applications

Wearable electronics has a very exciting but challenging market in this sector as wearable and fashionable clothes must make people want to wear them and to feel good when they do wear them.

The Philips/Levis ICD+ can be viewed as the first generation of smart clothing because they integrate mobile phones and music players, which try to enhance the 'organizer' functions of clothes. Until now, the closest wearable mobile communications has been Ericsson's Bluetooth headset. Nokia's prototype of the mobile snow jacket is an attempt to have devices such as the mobile phone fully incorporated within clothing. France Telecom has developed a phone coat equipped with an extra-flat (100 g) mobile telephone integrated in its lining. The keypad is placed on the sleeve of the jacket and a microphone is discretely placed on the collar. Infleon's MP3 jacket is an application in the area of infotainment, a combination of information and entertainment.

Cloth has always acted as an interface between the body and the external world. Wearables are no exception. They offer ample opportunities for the creation of intelligent clothing that perform functions according to the body's needs and requirements, and that adapts to the environment. Some of the functions are described as follows. Intelligent clothes can give reminders to people, identifying and memorising different objects to take with them, such as keys and wallets. They may perform temperature feedback and control mechanisms in a smart jacket, adjusting its interior temperature accordingly. Wearable medical devices can be integrated into a biosensor layer to monitor body conditions such as heartbeat, blood pressure and temperature. Another function is called the Virtual Doctor, which would assess and give advice on the overall health of the individual. In the future, such clothing may detect the user's feelings, moods, aches and pains, and so forth by monitoring brain activity and changing its colour, pattern, shape, and even its smell. However, fashion items worn by ordinary consumers require a high standard of quality and easy care. They should be washable in addition to being flexible and robust, and ideally foolproof. This is a real challenge to most electronic devices today.

1.3.4 Military and industrial applications

This is a very promising market where earlier penetration of the technology is expected to take place. In combat, soldiers must use their hands at all times to control weapons and machinery. Wearable devices to assist them would be very useful. The soldiers may be connected to navigation systems via wearable computers to guide them through difficult terrain and unknown areas. The systems may also let them know the positions of enemy and allied soldiers using satellite systems. Soldiers would also be in constant touch with their superiors. Others in nearby areas can be notified if a soldier falls down. Soldiers can look up on a stored database of information how to fix any damaged equipment and even how to apply first aid to injured soldiers. Other people, such miners and mountaineers, may need navigation and detection systems to guide them in avoiding dangerous areas and reaching safety.

Wearable electronic and photonics are likely to be expensive, initially only affordable by the military and by industrial sectors where performance is in great demand. The high costs of wearable technology could easily be outweighed by its efficiency and by the competitive edge it gives the user.

1.4 Implications of wearable technology

1.4.1 Economic impact

Wearable technology opens a door to many exciting applications and may lead to another technological revolution similar to the internet and mobile communication industries. The potential economic impact is enormous. It could lead to great opportunities for both the electronics and fashion/textile industries, each of which represents approximately 450 billion US dollars in world trade. The Venture Development Corporation estimated in 2003 that global market volume for smart fabrics and intelligent clothing, which includes wearable electronics and photonics, will reach 720 million US dollars in 2008, for an impressive annual growth rate of 18.8% between 2003 and 2008. The detailed estimation is given in Table 1.1.

1.4.2 Social and cultural factors

We are living in an exciting era and are feeling the great impact of technological advancement. In the past, the inventions of paper and the computer had profound influence on our society and culture, as well as on our lifestyle. It is expected that wearable technology will exert more influences in addition to those already made by portable electronic devices. Will social interaction increase or decrease when

Table 1.1 Market estimation for smart fabrics and intelligent clothing (Venture Development Corporation, 2003)

Market	$US × 1000		
	2003	2008	Annual increase rate (%)
Consumer sector	122 205	251 691	15.5
Professional and industrial sectors	150 884	388 086	20.8
Government	30 751	80 223	21.1
Total	303 840	720 000	18.8

most people possess clothing with wearable electronics and photonics? This may reduce the time to collect and communicate information, thus leaving more time for leisure and a social life.

The boundaries between science and engineering, which have traditionally been separate and distinct fields, have become blurred and the results of multidisciplinary and interdisciplinary research have been astonishing. Wearable electronics and photonics represent some of these results. This will have a profound influence on the future development of education and research. In the future, children may need to depend less on such abilities as reading and writing, if they are to be brought up in an environment where multimedia communication is made much easier for them. Meanwhile they may develop other abilities. Classes of the future may not take the present form, as people can learn things *in situ* with the aid of wearable computers and databases. The evidence indicates that face-to-face interactions reduce the levels of hormones involved in producing stress, fear and worry and increase levels of trust, bonding, attention and pleasure hormones. Hence, the technology of wearable electronics can be developed in such a way as to promote such face-to-face interactions, for instance, videophones.

In the future, people may not have as much privacy as they have today, as long as the problem of network security remains unsolved. A real threat is that people equipped with wearable electronics can outsmart other people who do not have such devices. What would happen if such technology falls into the wrong hands? A self-destructive trigger may provide some safeguard.

1.4.3 Health issues

The widespread use of mobile phones has led to a substantial amount of public concern over the possible adverse effects of electromagnetic waves on human brains. Many surveys and studies have been carried out, but no evidence has been found to support the view that mobile phones are harmful to human brains. Nevertheless, in order to avoid this possible hazard, new technologies employ an intelligent antenna to cut radiation by continually adjusting its characteristics to ensure that the power transmitted and received by mobile phones is directed away from the brain.

Another issue is the interference to normal operations of the human body by the wearable electronic devices. Humans have attained their current anatomy through a long process of natural evolution spanning tens of thousands of years. Current wearable computers need a head-mounted display as an output interface. Because the wearer's eyes always focus on the same spot on the screen, he/she may feel dizzy. Due attention will be paid to the biological and physiological aspects of humans when designing wearable products.

1.5 References

Bar-Cohen Y (2001), *Electroactive Polymer (EAP) Actuators as Artificial Muscles, Reality, Potential and Challenges*, SPIE Press, USA.

Koninklijke Philips Electronics NV (2000), *New Nomads*, 010 Publishers, Rotterdam, Netherlands.

Tao X M (2001), *Smart Fibres, Fabrics and Clothing*, Woodhead Publishing Ltd., Cambridge, UK.

Venture Development Corporation (2003), *Smart Fabrics and Intelligent Clothing*, Boston, USA.

2
Electrostatically generated nanofibres for wearable electronics

FRANK K. KO, AFAF EL-AUFY and HOA LAM
Drexel University, USA

ALAN G. MACDIARMID
University of Pennsylvania, USA

2.1 Introduction

Wearable electronics are electronic devices constantly worn by a person as unobstructively as clothing to provide intelligent assistance that augments memory, intellect, creativity, communication and physical senses.[1,2] Wearable electronics can be worn internally as implantable devices such as pacemakers and neuroprosthetics. Wearable electronics may be worn externally in the form of a ring, a badge, a wristwatch, eyeglasses, jewellery, shoes or clothing. The key considerations for wearable electronics are that they have to be robust, small, consume a small amount of power, and be comfortable to wear. Wearable electronics can function as sensors or as computers that consist of input, output and a motherboard made up of transistors and various interconnections. The wearable computer must be powered by lightweight batteries or fuel cells and must not be much of an additional burden to the wearer. Accordingly, the favourable ingredients for wearable electronics are lightweight, flexible and conductive materials. Conductive materials in fibrous form such as yarns and fabrics are preferred candidates for wearable electronics by serving as interconnects, functional devices and sensors.

Currently, most of the commercially available conductive yarns comprise a blend of non-conductive polymers with conductive particles such as carbon black, metallic particles, blends of polymeric fibres and continuous stainless steel fibres, stainless steel spun fibres and metal-clad aramid fibres.[3] Conductive yarns containing metallic components are prone to damage owing to excessive exposure to moisture and continuous fatigue during wearing and washing. The high stiffness of the metallic wires reduces the flexibility of the fabric, causing discomfort and restricting the mobility of the wearer. Commonly, the wearable electronic device is produced as a separate unit that can be attached to the garment. This allows the device to be removed from the clothes before washing, which is not convenient in

Table 2.1 Commercially available conductive yarns

Thread type	Conductivity (S cm^{-1})
BK 50/2 – Bekaert fibre technologies (steel/polyester spun yarn)	50
Bekintex – Bekaert fibre (continuous cold drawn stainless steel fibres)	1
Bekintex 15/2 – Bekaert fibre (stainless spun fibres)	1
VN 140 nyl/35 × 3 (nylon core wrapped with continuous stainless steel wires)	10
Aracon – DuPont (metal clad aramid fibre) Core: Kevlar Cladding metals: Ag, Ni, Cu, Au, Sn	0.001

some cases. Another method is to integrate the conductive yarns into a base fabric by means of embroidering, weaving, or knitting to generate a circuit pattern. This technique has recently been studied in wearable electronics and wearable computers.[3] Some of the commercially available conductive yarns, along with a brief description and their approximate resistance per unit length, are listed in Table 2.1. These fibres are available in diameters ranging from 100 μm to 12 μm.

The shortcomings of metal-based yarns can be overcome by the use of intrinsically conductive polymers (ICPs) or by a combination of ICPs with non-conductive polymers. ICPs, also known as synthetic metals, are organic polymers that possess the electrical, electronic, magnetic and optical properties of a metal, while retaining mechanical properties and processability. The properties of ICPs are intrinsic to a 'doped' form of the polymer.[4] Polymer-based devices are expected to have the combined advantages of low cost, control of material characteristics and flexible chemistry. Some ICP polymers are listed in Table 2.2, with polyacetylene, poly(3,4-ethylenedioxyphiothene)/poly(styrenesulphonate) (PEDT) and polyaniline (PANi) being the most commercially available and most extensively studied. They are being used in a wide range of applications, from actuators to rechargeable batteries.[3-6] These polymers and their blends are excellent candidates for wearable electronics because of their potential in the formation of fibres. One major concern with ICPs is their current-carrying capability over a long distance. The commercially available ICP fibres tend to be large and less flexible than textile fibres.

In order to improve the conductivity and other functional properties of wearable electronics, we are dedicated to exploring the potential use of nanofibres and nanofibre-based structures as wearable electronics. A significant reduction in the diameter of the fibres is expected to increase their flexibility greatly. The strength of the fibres and of the fibrous assemblies is also expected to increase. Owing to the

Table 2.2 Conductivity of ICPs

Polymer	Approx. conductivity (S cm^{-1})	Polymer	Approx. conductivity (S cm^{-1})
Polyacetylene	10 000[a]	Polythienylenevinylene	2700[a]
Polypyrrole	500–7500	Polyphenylene	1000
Polythiophene	1000	Polyisothianaphthene	50
Poly(3-alkylthiophene)	1000–10 000[a]	Polyazulene	1
Polyphenylene sulphide	500	Polyfuran	100
Polyphenylenevinylene	10 000[a]	Polyaniline	200[a]

[a] Conductivity of oriented polymer.

effect of confinement, unusual properties of current conduction are expected to emerge when the size of the fibres is reduced below a certain critical thickness/diameter to the nanoscale.[7,8] Several processes can be used that have the potential to make fibres in the nanoscale. These processes include the template method,[9] the island in the sea technology[10] and the electrostatic spinning process. Of these, the electrostatic spinning process is preferred because of its simplicity and its potential to scale into a continuous process. The electrospinning process allows the nanofibres (diameter less than 100 nm) of organic polymers to be controllably and reproducibly fabricated such that, in one given preparation, all fibres will have a diameter of less than 100 nm. The objective of this chapter is to introduce an electrostatic spinning process for the formation of nanofibres of pure electronic polymers (in their semiconducting and metallic regimes) and/or their blends in conventional organic polymers for the purpose of ascertaining their applicability in the fabrication of nanoelectronic devices for wearable electronics.

2.2 Electrospinning process

2.2.1 Background

The electrostatic generation of ultrafine fibres, a process called electrospinning, has been practised since the 1930s.[11] This technique has been rediscovered for applications such as high-performance filters[12,13] and for scaffolds in tissue engineering[14] that utilise the high surface area (10 m^2 g^{-1}) unique to these fibres. In this non-mechanical, electrostatic technique, a high electric field is generated between a polymer fluid (contained in a glass syringe with a capillary tip) and a metallic collection screen. When the voltage reaches a critical value, the charge overcomes the surface tension on the deformed drop of the suspended polymer solution that is formed on the tip of the syringe, and a jet is produced. During its passage to the collection screen, the electrically charged jet undergoes a series of electrically induced bending instabilities that results in the hyperstretching of the

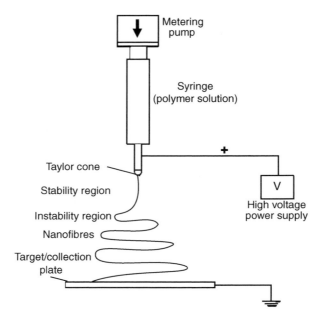

2.1 Schematic of the electrospinning process.

jet. This stretching process is accompanied by rapid evaporation of the solvent molecules that reduces the diameter of the jet to a cone-shaped volume called the envelope cone. The dry fibres are accumulated on the surface of the collection screen, resulting in a non-woven mesh of fibres with diameters between nanometres and micrometres. The process can be adjusted to control the diameter of the fibre by varying the strength of the electrical field and concentration of the polymer solution, while the duration of electrospinning controls the thickness of the deposition of fibres.[11] A schematic drawing of the electrospinning process is shown in Fig. 2.1. Considering the fact that the rate of electrochemical reactions is proportional to the surface area of the electrode, it is of interest to explore the merits of the high surface to volume ratio of electrospun fibres in order to develop porous polymeric electrodes. Reneker and Chun[15] have reported that polyaniline fibres can be successfully electrospun from sulphuric acid into a coagulation bath. Another way of producing electrically conductive nanofibres is by the pyrolysis of electrospun polyacrylonitrile nanofibres into carbon nanofibres.

2.2.2 Controlling the diameter of the fibre

In order to generate fibres in the nanoscale consistently and reproducibly, it is important to understand the parameters affecting the diameter of the fibres by the electrospinning process. Although the process of electrospinning has been known for over half a century, current understanding of the process and those parameters is very limited. Many processing parameters influence the spinnability and physical properties of nanofibres, including the strength of the electric field, the concentration of the polymers, spinning distance, viscosity of the polymer, and so forth.[16] Experimental evidence has shown that the diameter of the fibre produced by electrospinning is influenced by the concentration of the polymer and by molecular conformation.[17] To establish a processing index for controlling the diameter of fibres, the Berry number (Be), a dimensionless parameter,[18] was used:

$$Be = [\eta]*C$$

where $[\eta]$ is the intrinsic viscosity of the polymer and the ratio of the specific viscosity to the concentration at an infinite dilution, and C is the concentration of the polymer solution. Intrinsic viscosity is also closely related to the molecular weight of the polymer. It has been found that the degree of the entanglement of polymer chains in a solution can be described by Be.[18] In very diluted solutions, when the Be is less than unity, the molecules of the polymer are sparsely distributed in the solution. There is a low probability of individual molecules becoming entangled in each other. At a Be of greater than unity and as the concentration of the polymers increases, the level of molecular entanglement increases, resulting in more favourable conditions for the formation of fibres.

The relationships between the concentration and the average fibre diameter (AFD) of a poly(L-lactic acid) (PLA) of different molecular weights in chloroform are shown in Fig. 2.2. It can be seen that the diameter of the fibre increases as the concentration of the polymer increases. The diameter of the fibre increases at a greater rate at higher molecular weights. Accordingly, one can tailor the diameter of the fibre through the proper selection of polymer molecular weight and polymer concentration.

Expressing the diameter of a fibre as a function of Be, as shown in Fig. 2.3, a pattern emerges indicating four regions of Be–diameter relationships. Region I, where $Be < 1$, is characterised by a very diluted polymer solution with molecular chains that barely touch each other. This makes it almost impossible to form fibres by the electrospinning of such a solution, since the chains are not entangled enough to form a continuous fibre and the effect of surface tension will make the extended conformation of a single molecule unstable. As a result, only polymer droplets are formed. In region II, where $1 < Be < 3$, AFD increases slowly with Be from ~100 to ~500 nm. In this region, the degree of molecular entanglement is just sufficient for fibres to form. The coiled macromolecules of the dissolved polymer are transformed by the elongational flow of the polymer jet into oriented molecular assemblies with some level of inter- and intra-molecular entanglement. These entangled networks persist as the fibre solidifies. In this region, some bead formations are observed as a result of the relaxation of the polymers and the effect of surface tension. In region III, where $3 < Be < 4$, AFD increases rapidly with Be, from ~1700 to ~2800 nm. In this region the entanglement of the molecular chain becomes more intensive, contributing to an increase in the viscosity of the polymer. Because of the intense level of molecular entanglement, a stronger

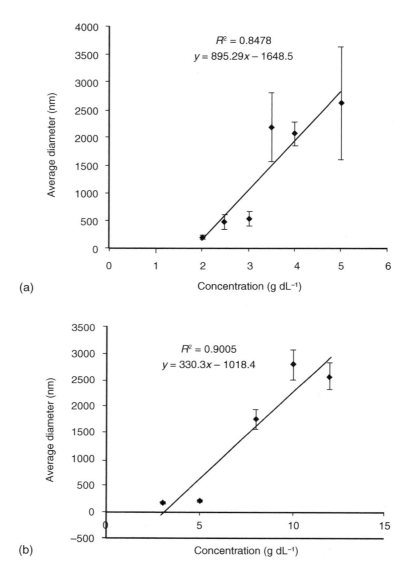

2.2 Relationships of fibre diameter to concentrations at different molecular weights (MW). (a) MW = 300 000, (b) MW = 200 000.

electric field is needed for fibres to form by electrospinning. In region IV, where $Be > 4$, the AFD is less dependent on Be. With a high degree of inter- and intramolecular chain entanglement, other processing parameters such as the strength of the electric field and spinning distance become dominant factors that affect the diameter of the fibre. A schematic illustration of the four Berry regions is shown in Table 2.3.

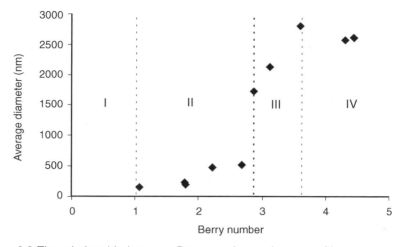

2.3 The relationship between Berry number and average fibre diameter.

Table 2.3 Schematic of polymer chain conformation and fibre morphology corresponding to the four regions of the Berry number

	Region I	Region II	Region III	Region IV
Berry number	$Be < 1$	$1 < Be < 2.7$	$2.7 < Be < 3.6$	$Be > 3.6$
Polymer chain conformation in solution				
Fibre morphology				
Average fibre diameter (nm)	(Only droplets formed)	~100–500	1700–2800	~2500–3000

2.2.3 Formation of yarns and fabrics

Electrospun fibres can be assembled by direct fibre-to-fabric formation to create a non-woven fabric or by the creation of a linear assembly or a yarn. From the yarns, a fabric can be woven, knitted or braided. The linear fibre assemblies can be aligned mechanically or by controlling the electrostatic field. Alternatively, a self-assembled continuous yarn can be formed during electrospinning by properly designing the ground electrode. The rotary electrospinning method for the formation of aligned nanofibres and the self-assembled continuous yarn process are described here.

Rotary electrospinning

A schematic illustration of rotary electrospinning is shown in Fig. 2.4. The rotary electrospinning system consists of a polymer reservoir, a spinneret, a power supply and a variable speed rotating fibre-collecting disk. As the polymer jet is drawn from the polymer cone, it travels toward the spinning disk. The solidified fibres are deposited on the collector as the solvent evaporates. Since the disk is rotating, the deposited fibres are placed in the direction of the rotation, forming rings of aligned fibres on the disk. The fibres are subsequently collected in batches or continuously collected into well-aligned bundles of yarn.

Electrospinning of self-assembled yarn

Self-assembled yarn can be produced in a continuous length with appropriate control of electrospinning parameters and conditions. In this process, the self-assembled yarn is initiated by limiting the footprint of the deposition of fibre into a small area on the target. The fibres are allowed to build on top of each other until a branched tree-like structure is formed. Once a sufficient length of yarn has

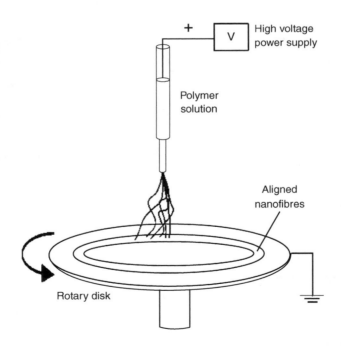

2.4 Schematic drawing of the rotary disk electrospinning process.

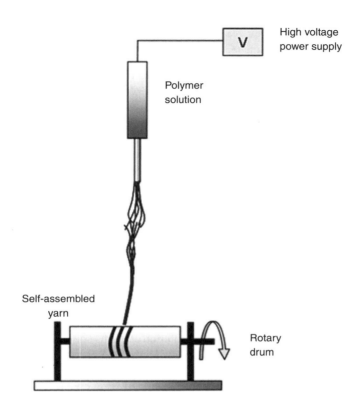

2.5 Schematic of the continuous electrospinning of self-assembled yarn.

formed, the accumulated fibres attach themselves to the branches and continue to build up. A device such as a rotating drum can be used to spool up in a continuous length the self-assembled yarn that has been produced, as shown in Fig. 2.5. This method produces bundles of partially aligned nanofibre yarn in a continuous length.

2.3 Electroactive nanofibres

2.3.1 Inherently conductive polymers and blends

ICPs and blends of ICPs with other fibre-forming polymers can be electrospun into linear and planar assemblies. We have previously reported[19] the fabrication of the first conducting polymer fibres (diameter ~950 to 2100 nm) of polyaniline doped with d,l-camphorsulphonic acid (PANi.HCSA) as a blend in polyethylene oxide (PEO). It was discovered that an electronic polymer such as polyaniline, which might have been expected to be more susceptible to degradation than most conventional organic polymers, survived without observable chemical or physical change following the 25 kV electrospinning–fabrication process in air at room temperature. The fibre diameter morphology of the polymer-blend electrospun fibres was examined using a scanning electron microscope (SEM). According to the SEM micrographs for the different blends of PANi.HCSA/PEO fibres that were investigated, these fibres showed a similar diameter and morphology as the PEO fibres. For example, the SEM micrograph of the non-woven mat of electrospun fibres from a 2 wt% PANi.HCSA/2 wt% PEO solution, shown in Fig. 2.6, revealed that the fibres in the non-woven mat ranged from between 950 nm and 1.9 μm (average fibre diameter, 1.6 μm) and had a generally uniform thickness. Again, the electrospun fibres were distributed randomly in the non-woven mat. From the SEM micrographs of all of the different polyaniline/PEO blends that were electrospun at an electrical field strength of 1 kV cm^{-1}, it appears that the addition of PANi.HCSA to the PEO solution had little effect on the diameter of the electrospun fibre. The PANi.HCSA/PEO electrospun fibres that were produced showed no evidence of birefringence, thus indicating that the polymer chains in the fibre were not oriented with respect to the axis of the fibre.

The conductivity of the electrospun PANi.HCSA/PEO fibres and the cast film on a microscope glass slide was measured using the four-point probe method.[20] The thickness of the non-woven fibre mat and the cast films were measured using a digital micrometer (Mitutoyo) with a resolution of 1 μm. The current was applied between the outer electrodes using a Princeton Analytical Research 363 potentiostat/galvanostat, and the resulting potential drop between the inner electrodes was measured with a Keithley 169 multimeter. Figure 2.7 shows the room temperature conductivity of the PANi.HCSA/PEO electrospun fibres and cast films at various

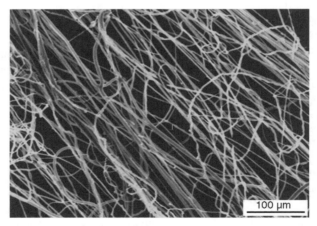

2.6 Nanofibre blend of PANi.HCSA fabricated from 2 wt% PANi.HCSA and 2 wt% PEO in chloroform solution at 25 kV (anode/cathode separation, 25 cm).

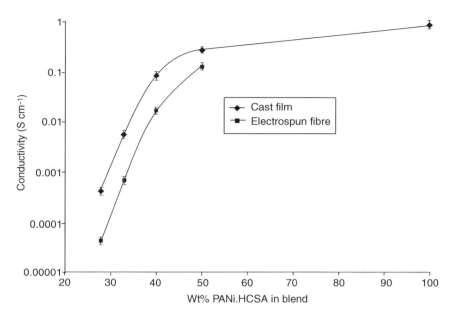

2.7 Electrical conductivity of the PANi.HCSA/PEO blend electrospun fibres and cast films prepared from the same solution (the unit for conductivity on the ordinate is S cm^{-1} based on a four probe measurement).

ratios of polyaniline and polyethylene oxide in the blend. This graph demonstrates that the conductivity of the electrospun fibres in the non-woven mat is significantly lower than that for a cast film at the same concentration of polyaniline. This is not an unexpected result, as the four-point probe method measures the volume resistivity from which the conductivity can then be calculated and not that of an individual fibre. It must be noted that obtaining the conductivity of the non-woven mat was considerably more difficult than measuring the conductivity of the cast film, owing to the difficulty in obtaining an accurate measurement of thickness on the highly compressible non-woven mat using the micrometer. As can be seen from the SEM micrographs of the electrospun fibres (Fig. 2.6), the non-woven mat is highly porous and therefore the 'fill factor' of the polyaniline fibres is less than that of a cast film. However, it is reasonable to expect that the conductivity of an individual electrospun fibre will be higher than that of the non-woven mat and, in fact, should be approximately equal to the conductivity of the cast film.

Since the submicrometre fibres (500–1600 nm) obtained in our initial work (Fig. 2.6) were not classifiable as true nanofibres, we turned our attention to the consistent and reproducible fabrication of the true nanofibres (diameter <100 nm) of an organic polymer. This was accomplished using an 8 wt% solution of polystyrene (MW 212 400, Aldrich Co) in tetrahydrofuran (THF) (glass pipette orifice, 1.2 mm) at a potential of 20 kV between the anode and cathode, which were separated by 30 cm. The fibres were collected as a mat on an aluminum target and found by SEM to have the following diameter characteristics: average, 43.1 nm; maximum, 55.0 nm; minimum, 26.9 nm. Other studies involved polystyrene-produced fibres whose diameters were consistently less than 100 nm. For example, another sample of polystyrene had the following fibre diameter characteristics: average, 30.5 nm; maximum, 44.8 nm; minimum, 16.0 nm. It should be noted that these data represent a decrease (of approximately two orders of magnitude) in fibre diameter as compared to those obtained in our earlier studies (Fig. 2.6). It should also be noted that the above 16 nm fibre is ~ 30 polystyrene molecules wide. Dimensions of this size can be expected greatly to affect the kinetics, as well as possibly the thermodynamics, of the polymer. It is also of interest to note that a 16 nm fibre, such as the one mentioned above, lies well within the ~ 4–30 nm range of the diameter of multiwalled carbon nanotubes.[21]

Using a previously observed method for producing polyaniline fibres,[19] we prepared highly conducting sulphuric acid-doped polyaniline fibres (average, 139 nm; maximum, 275 nm; minimum, 96 nm) by placing a ~ 20 wt% solution of polyaniline (Versicon™ Allied Signal) in 98% sulphuric acid in a glass pipette. The tip of the pipette was ~ 3 cm above the surface of a copper cathode immersed in pure water at a potential difference of 5 kV. The fibres collected in or on the surface of the water. As expected, the conductivity of a single fibre was ~ 0.1 S cm^{-1}, since partial de-doping of the fibre occurred in the water cathode.

It is relatively easy to prepare conducting blends of PANi.HCSA in a variety of different conventional polymers such as PEO, polystyrene, polyacrylonitrile, and so forth. For example, ~ 20 wt% blends of PANi.HCSA in polystyrene (MW 114 200) are obtained by electrospinning in a chloroform solution. Their fibre diameter characteristics are: average, 85.8 nm; maximum, 100.0 nm; minimum 72.0 nm. These fibres are sufficiently electrically conductive that their SEMs may be recorded without the necessity of applying a gold coating. Separate, individual nanofibres can be collected and examined if so desired. An appropriate substrate – a glass slide, silicon wafer, or loop of copper wire, etc. – is held between the anode and cathode at a position close to the cathode for a few seconds to collect individual fibres. The current/voltage (I/V) curves are given in Fig. 2.8 for a single 419 nm diameter fibre (fibre 1) of a blend of 50 wt% PANi.HCSA, and polyethylene oxide is collected on a silicon wafer coated with a thin layer of SiO_2. Two gold electrodes separated by 60.3 μm are deposited on the fibre after its deposition on the substrate. The conductivity (two probes) of fibre 1 (diameter ~ 600 nm) is ~ 10^{-1} S cm^{-1}. The conductivity (two probe) of fibre 2 (diameter ~ 419 nm) is ~ 10^{-1} S cm^{-1}. Non-linear I/V curves may be obtained from some polyaniline samples, possibly caused by the presence of defect sites induced by imperfections or impurities in the polyaniline. Such imperfections are expected to be more apparent in thin fibres, since there are fewer molecular pathways by which charge carriers can by-pass the defect sites.

The (reversible) conductivity/temperature relationship between 295 and 77 K

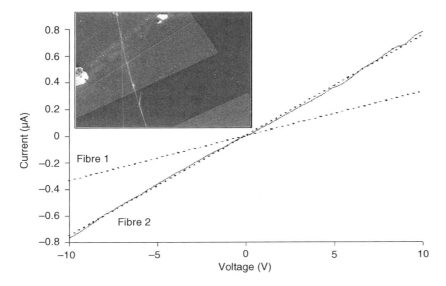

2.8 Current/voltage curves of 50 wt% PANi.HCSA/PEO blend nanofibres. The inset shows a fibre spun from a blend of 50 wt% PANi.HCSA and polyethylene oxide collected on a silicon wafer coated with a thin layer of SiO_2.

for a single 1320 nm fibre containing 72 wt% PANi.HCSA in PEO spun from chloroform solution is given in Fig. 2.9. To minimise the effects of heating, the applied voltage was held constant at 10 mV, at which value the current is very small. The conductivity (~ 33 S cm^{-1} at 295 K) was unexpectedly large for a blend since the conductivity of a spun film of the pure polymer cast from chloroform solution is only ~ 10^{-1} S cm^{-1}.[22] This suggests that there may be a significant alignment of polymer chains in the fibre.[23]

Poly(3,4-ethylenedioxythiophene)/poly(styrenesulphonate) (Eleflex-2000) and PEDT are popular electronic polymers used for electronic components such as light-emitting diodes (LED). It has been demonstrated in our laboratory that PEDT can be mixed with various polymers to form conductive nanofibres by adding an appropriate amount of PEDT as a percentage of weight to a solvent such as N,N-dimethylformamide (DMF) and stirred magnetically for 10–15 minutes. The mixture is then added to a polymer such as polyacrylonitrile (PAN) and stirred for an additional 20 minutes at 60°C.[24]

Figure 2.10 shows an SEM image of the PEDT/PAN electrospun fibres from a combination of 20 wt% PEDT and 8 wt% PAN. A high level of fibre alignment was achieved by the yarn-self-assembly process during electrospinning.

In Fig. 2.11, the diameters of the PEDT/PAN fibres are shown for various weight percent of PAN and PEDT. As expected, the diameter of the fibre decreases

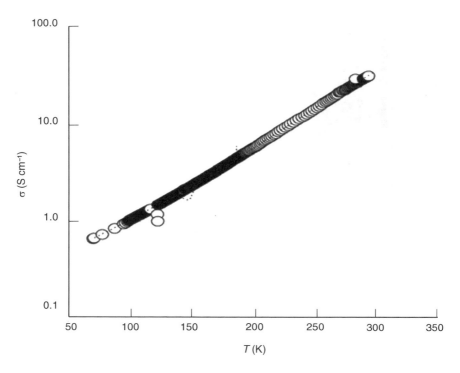

2.9 Conductivity/temperature relationship for a 72 wt% blend fibre of PANi.HCSA in PEO.

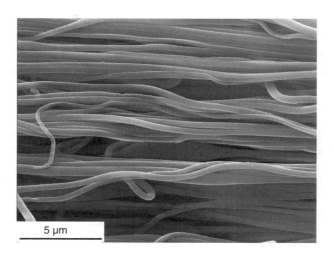

2.10 SEM image of aligned PEDT/PAN nanofibres by electrospinning.

Electrostatically generated nanofibres for wearable electronics

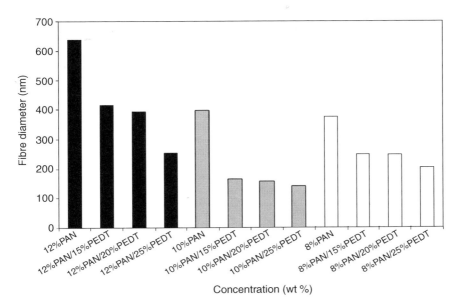

2.11 Diameter of a fibre at various PEDT/PAN concentrations.

as the concentration of polymer-spinning dope decreases. It is of interest to note that, for the same level of PAN concentration, the addition of PEDT led to a reduction in the diameter of the fibre.

The current–voltage relationship of the PEDT/PAN fibre webs was characterised by the four-probe method. This was done by electrospinning the PEDT/PAN polymer directly on a silicon wafer. The I/V curves of the Si wafer with and without the fibres are shown in Fig. 2.12, from which the resistance of the wafer and wafer/fibre assemblies were calculated. By considering the Si wafer and the fibre mat to be resistors in a parallel connection, the resistance of the electrospun fibres can be calculated by applying the following relationships:

$$V = I * R$$

$$\rho = R * A / L$$

where V is the applied electrical potential, I is the current, R is the resistance of the material, A is the cross-sectional area perpendicular to the direction of the current, L is the distance between the two points at which the voltage is measured and ρ is the resistivity of the material (in Ω-cm). The conductivity (S cm^{-1}) of the material can be calculated from the reciprocal of the resistivity.

From Fig. 2.13, it is observed that the diameter of a fibre does indeed play an important role in the conductivity of the fibre; smaller fibres tend to have higher electrical conductivity. As shown in Fig. 2.14, the increase in the concentration of PEDT in the fibril matrix also resulted in an increase in the electrical

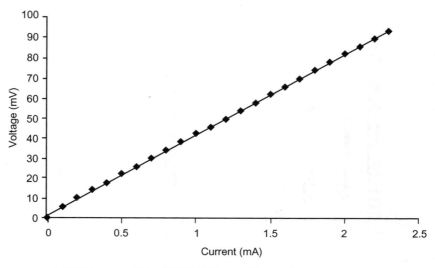

2.12 I/V curve of the PEDT/PAN nanofibre webs.

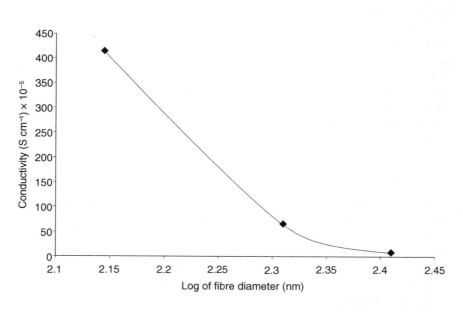

2.13 Effect of fibre diameter on electrical conductivity of PEDT/PAN fibre webs.

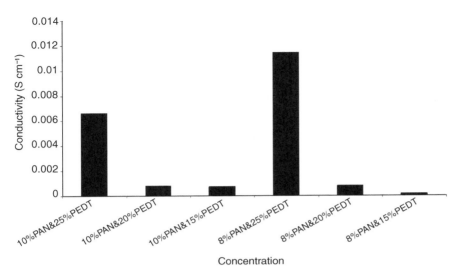

2.14 Effect of concentration of PEDT on the conductivity of the fibres.

conductivity of the yarns. The measured level of electrical conductivity of the PEDT/PAN yarn was 0.001–0.012 S cm^{-1}.

2.3.2 Nanocomposites

Carbon black, metallic nanoparticles, graphite nanoplatelet and carbon nanotubes can be mixed with conductive and nanoconductive polymers to form spinning dopes for the co-electrospinning of conductive nanocomposite fibrils. Taking advantage of the high level of electrical conductivity and superior mechanical properties of carbon nanotubes,[25] the nanocomposite fibril concept will be demonstrated here using single-wall carbon nanotubes (SWNTs) as the filler.

SWNTs produced by the high pressure disproportionation of carbon monoxide (HiPco) method with an average diameter of 1.5 nm and a length of 1–2 μm were mixed in a solution of PAN/DMF to form a spinning dope for co-electrospinning. The SWNTs were magnetically stirred in the DMF for 12 h. Polyvinylpyrrolidone (PVP) was used as a surfactant to wrap the nanotubes and prevent them from re-aggregating or agglomerating. After adding the PVP to the SWNT/DMF, the suspension was sonicated for 20 min. An appropriate amount of PEDT was added to the mixture and the mixture was magnetically stirred for 15 min. Finally, a measured amount of PAN was added to the previous mixture and stirred for an additional 25 min at ~ 90°C.

The co-electrospinning of SWNT/PEDT/PAN nanofibres was successfully demonstrated with various concentrations of SWNTs, ranging from 0.2, 0.5 and 0.8 to 1 wt% in a 20 wt% PEDT/8 wt% PAN solution. To account for the increase

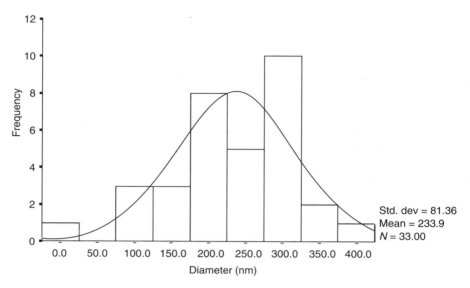

2.15 Fibre diameter distribution of the electrospun SWNT/PEDT fibrils.

in viscosity caused by the addition of SWNTs, the concentration of polymers for the spinning dope was adjusted accordingly. Thus, for spinning dopes containing 3 and 5 wt% of SWNTs, the concentration of polymers was reduced to 10 wt% PEDT/5 wt% PAN. Figure 2.15 shows the diameter distribution of the electrospun composite nanofibres. An average fibre diameter of 230 nm was obtained from the sample.

To verify the presence of SWNT in the nanofibrils, the composite fibrils were characterised by transmission electron microscopy (TEM) and Raman spectroscopy. Figure 2.16 (a) shows the crystalline structure of the PEDT embedded in the PAN fibre with an average diameter of 200 nm. Figure 2.16 (b) illustrates the alignment of SWNT along the axis of the fibre, showing evidence that SWNT are incorporated in the nanocomposite fibrils. The diameter of the SWNT was measured to be approximately 1.2 nm, as indicated in Fig. 2.17.

The inclusion of SWNTs in the PEDT/PAN matrix fibril was further confirmed using Raman microspectroscopy (Renishaw 1000 Raman Microspectrometer) with a diode laser (780 nm excitation wavelength, 12 W cm^{-2}). Figure 2.18 shows the Raman spectra of pure PAN fibres and 0.5–1.0 wt% SWNT/PEDT/PAN fibres. The typical peaks of SWNT are the radial breathing mode (RBM) in the 100–275 cm^{-1} range, and the tangential (stretching) mode in the 1500–1600 cm^{-1} range. These peaks are evident in the SWNT/PEDT/PAN fibrils, but absent in the pristine PAN fibrils. The diameter of the SWNT can be estimated from the RBM peaks using the equation: $\omega_R \sim 224$ (cm^{-1} nm)/d, where ω_R is the RBM frequency and d is the diameter of the tube in nanometers. The presence of at least 5 RBM peaks is observed in the range from 153–267 cm^{-1}, which corresponds to diameters of tubes in the range of 0.8–1.5 nm.

Electrostatically generated nanofibres for wearable electronics 31

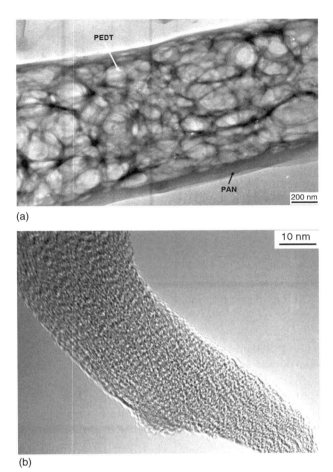

2.16 (a) TEM image of PEDT/PAN nanofibre; and (b) TEM image showing the alignment of SWNT in the PEDT/PAN nanofibres.

The electrical conductivity of the SWNT-filled nanocomposites was characterised by the four-probe method as described in Section 2.3.1. Figure 2.19 shows that the conductivity of the fibrils increases as the weight percent of PEDT increases. The introduction of a few weight percent of SWNT leads to a significant increase in conductivity.

The effect of adding SWNTs in the PEDT/PAN matrix on the electrical conductivity of the fibrils is shown in Fig. 2.19. The presence of a few percent of SWNTs results in a doubling of the conductivity of the fibres to the level of 0.03 S cm^{-1}.

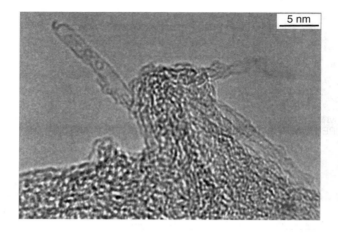

2.17 TEM image showing SWNT protruding from the fracture surfaces of the fibre.

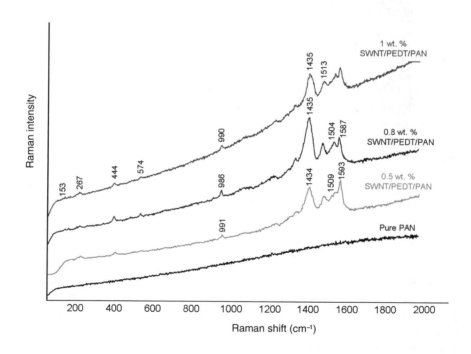

2.18 Raman spectra of composite nanofibrils of pure PAN and PEDT/PAN with various wt% of SWNTs using an excitation wavelength of 780 nm.

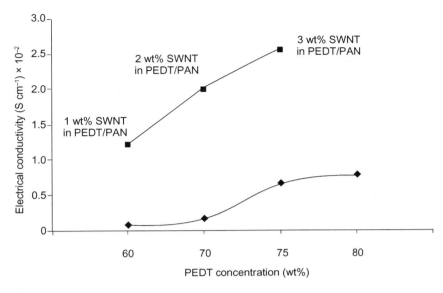

2.19 The electrical conductivity of PEDT/PAN fibrils at various levels of PEDT concentration with and without the presence of SWNTs.

2.3.3 Pyrolysis and coating of nanofibres

Electrospun fibres from organic polymers can be used as a precursor for converting to conductive nanocarbon fibres or simply used as a template for coating with a conductive polymer or metal.

Pyrolysis

As previously reported,[26] polyacrylonitrile fibres may be thermally converted to carbon nanofibres with some shrinkage. We have similarly converted a polyacrylonitrile fibre (diameter 750 nm) to a carbon fibre by first heating it in air at 200°C for 20 min, followed by heating at 800°C for 2 h under nitrogen. A current/voltage curve was obtained for a carbon fibre 600 nm in diameter. The controlled conversion of organic electrospun polymer fibres presents interesting opportunities for the fabrication of a variety of carbon nanofibres.

Electrodeless coating

The large surface-to-volume ratio offered by nanofibres makes them excellent substrates for the fabrication of coaxial nanofibres consisting of superimposed layers of different materials. Catalysts and electronically active materials can be deposited by chemical, electrochemical, solvent, chemical vapour or other means, for use in nanoelectronic junctions and devices. It was found that polyacrylonitrile nanofibres can be easily and evenly coated with a 20–25 nm layer of conducting polypyrrole by immersion in an aqueous solution of polymerising polypyrrole.[27, 28] Analogously, we have found that the electrodeless deposition of metals can also be performed. Polyacrylonitrile fibres, for example, can be evenly coated with gold by treating them with a solution of AuS_2O_3 and ascorbic acid.

2.4 Ultra-low dielectric constant of nanocomposite fibrous film

The dielectric constant is an important electrical property of materials. It can be defined as the ratio of forces between two charges *in vacuo* to that in the medium. Alternatively, dielectric constant can be defined as the ratio of the capacity of a condenser to the capacity of the condenser *in vacuo*. The dielectric effects are the results of the polarisation of the medium between two charges when the medium is subjected to an electric field. Dielectric constant provides an indication of the relative speed that an electrical signal will travel in the medium. A medium with a low dielectric constant will result in a higher signal propagation speed since the speed of a signal is inversely proportional to the square root of the dielectric constant.[29]

Materials with a high dielectric constant are suitable for application in capacitors,[30, 31] whereas materials with a low dielectric constant are important in microelectronic and integrated circuit applications. Major advances have been made in fast and powerful microelectronic devices through miniaturisation. As the devices are getting smaller, the intermetal dielectric (IMD) constant must be low in order to reduce the resistance–capacitance (RC) time delay of the interconnects and the cross talk between metal lines. It is well known that the benefit to be derived from low-dielectric constant materials is far greater than from increasing the conductivity of the metal interconnects.

Accordingly, many laboratories are actively searching for low dielectric constant, or low K, materials. The systems that have been proposed and tested include fluorinated silica glass,[32, 33] amorphous C:F,[34] air gap formation,[35] non-fluorinated polymers,[36] inorganic–organic hybrids, porous polymer (methylsilsesquioxane), porous silica materials (Aerogel, dried by the supercritical method; or Xerogel, dried by the evaporation of solvent at ambient temperature), polyimide nanofoams,[29] Parylene-F-like film,[37] polymeric, poly(arylethers) (PAEs) (2.7–1.8, 40% porosity),[29] and fluorinated silica xerogel film.[32] In addition to low K requirement, a major challenge is to obtain materials with a low dielectric constant that have adequate mechanical strength, chemical and thermal properties, and that are capable of being integrated into manufacturing integration.

Currently, the processes being used to produce low K materials include spin casting on glass[38] and fluorinated SiO_2 ($k > 3$); nanoporous silica; crack-free

Table 2.4 State-of-the-art low K materials

Process	Structure/trade name	Manufacturer	Dielectric constant
CVD	HSQ	DOW	3.6
CVD	High K OSG	Applied Materials	3.1
CVD	Low K OSG	Applied Materials	2.7
SOD	HSQ	DOW	2.6
SOD	Porous silk	DOW	2.6
SOD	Nanoglass	Allied Signal	2.5
SOD	PTFE	Gore	2.1

Xerogel film (1.3~2.2)[38, 39] produced according to a law of logarithmic mixture (Lichtenecker's rule); and chemical vapour deposition (plasma-enhanced). A summary of the properties of these low K materials are shown in Table 2.4.

An examination of the state-of-the-art low K materials shows that there is a general trend to use porous materials, presumably by taking advantage of the fundamental fact that the lowest dielectric constant of one is only attainable in a vacuum. A common feature of these low K materials is that they tend to be rigid systems and many of them are brittle although they must have enough strength to endure the handling involved during processing.

We report a new class of nanofibre/nanocomposite-based ultra-low K materials produced by the electrospinning process. Through the use of ultra-fine fibres, an extremely high level of pore volume and pore surfaces can be achieved while maintaining a high level of areal coverage. The fibrous nature of the nanofibrous assembly lends itself to a flexible and conformable structure. The mechanical, chemical and thermal properties of the nanofibrous assembly can be tailored by the introduction of a second and/or third phase such as nanoparticles, nanoplatelets and nanotubes. These nanofibre and nanocomposite systems can be produced by a simple, non-mechanical process through electrostatic spinning.

To illustrate the concept of nanofibrous low K materials, PAN solutions with N,N,-dimethylformamide (DMF) solvent were prepared with the compositions shown in Table 2.5. The spinning solutions were prepared by mixing PAN with DMF and sonicating the mixture for 20 min. Sonication was carried out in 5-min intervals in order to keep the solution from overheating. A magnetic stirrer was used to distribute uniformly the heat generated during sonication. For the composite solution sample, the added filler was first mixed with DMF and sonicated for 20 min before being mixed with PAN to form the spinning solution, following the same procedure outlined before. Iron oxide (Fe_2O_3) nanoparticles, graphite nanoplatelets (GNP) and carbon nanotubes (CNT) were used as fillers to form nanocomposite fibres. Following the process described above, an electrostatic field of 20 kV was applied over a distance of 15 cm between the tip of the syringe needle and a 75 mm × 75 mm (3" × 3") sample ground plate covered with

Table 2.5 Composition of PAN/DMF solutions for electrospinning

DMF volume: 30 ml
PAN/DMF: 7 wt%

DMF (g)	PAN (g)	Filler (g)	Wt% filler in PAN
26.5	1.855	0.019	1.0
26.5	1.855	0.036	2.0
26.5	1.855	0.056	3.0
26.5	1.855	0.074	4.0

Table 2.6 Dielectric constants of various nanofibrous materials measured at 100 kHz

Material	Frequency (Hz)	Dielectric constant
PAN-(A)	100k	1.46
PAN-(B)	100k	1.24
PAN-FE-1%	100k	1.16
PAN-FE-2%	100k	1.28
PAN-FE-4%-A	100k	1.14
PAN-FE-4%-B	100k	1.16
PAN-FE-8%	100k	1.08
PAN/GNP-2%-A	100k	1.18
PAN/GNP-2%-B	100k	1.16
PAN/GNP-4%-A	100k	1.07
PAN/GNP-4%-B	100k	1.15

aluminum foil. For the purposes of comparison, pristine polymer film and composite films were also prepared for subsequent characterisation of dielectric properties. A detailed description of the process has been reported elsewhere.[29]

Measurements of the dielectric constant of the nanofibrous structures were made with an Agilent 4263B LCR Meter with a 16451B Dielectric Test Fixture.[41] Electrode B of 16451B was used with a required sample diameter of between 10 and 50 mm. In order to facilitate a comparison of the test results, circular specimens were prepared with a nominal diameter of 12.5 mm. The thickness of the specimens was determined by taking an average of three measurements using a micrometer. All of the measurements were made by setting the LCR meter on a bias DC voltage of 1000 mV over the available frequency range of 1, 10 and 100 kHz. A non-contact measurement procedure or air-gap method was used for all the measurements of the dielectric constant. By measuring the capacitance of two measurements, one in which a sample was used and the other in which it was not used, this method eliminates the concern over the possible existence of gaps of

Electrostatically generated nanofibres for wearable electronics 37

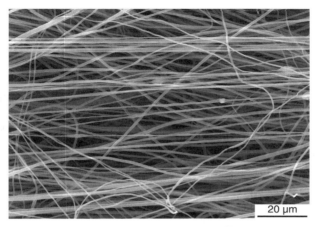

2.20 SEM image of an electrospun nanofibres mat.

air between the electrode and test sample. The dielectric constants of the nanofibrous membranes were measured at frequencies ranging from 1 to 100 kHz. The dielectric constants of various nanofibrous materials measured at 100 kHz are summarised in Table 2.6. The results show that the measured dielectric constant is between 1.1–1.46 for samples spun at 100 kHz to 20 kV. By comparison, the spun PAN has about a 39% lower dielectric constant than the cast PAN, and the composite-spun PAN has a further reduction of 12% in comparison to the spun PAN. This super-low dielectric constant associated with the nanofibrous structures may be attributed to the high level of porosity of the nanofibrous structures resulting from a tortuous three dimensional interconnected fibrous network.

The porosity of the spun fibre mats was estimated to be in the range of 70–90%. The intralayer porosity is illustrated in the SEM image shown in Fig. 2.20. This combination of high porosity, flexibility, and toughness is quite usual for nanofibrous membranes.

With further development, this family of nanofibrous materials promises to be able to meet a future demand for low K for microelectronic devices. The mechanical strength of the fibre and other requirements may be enhanced or increased by the incorporation of nanoparticles.

2.5 Conclusions

Inherently conductive polymers, polymer blends of ICP and their composites in nanofibre form and in the form of nanofibrous assemblies are promising candidates for wearable electronics. Taking advantage of the nanoscale effect, these polymer-based nanowires form the structural backbone of interconnects, functional devices, sensors, lightweight batteries for high-performance uniforms for soldiers, as well as for stylish and comfortable garments.

Electrospinning is a simple and non-mechanical process for the production of nanofibres. In order to transit this seemingly uncomplicated fibre-spinning process from a curiosity of the laboratory to a robust manufacturing process, there is a need to understand systematically the processing parameters. As a first step toward this goal, complementing ongoing electrohydrodynamic modelling,[40–46] a processing index was introduced as a means of relating the diameters of fibres to the Berry number. To convert these nanoscale fibres to clothing assemblies and integrated circuits, methods for the formation of yarn and fabric using the electrospinning process were introduced.

To demonstrate the broad range of material design concepts, the electronic properties of electrospun polyaniline, polyaniline/PEO blend, PEDT/PAN, PEDT/PAN/SWNT and PLA/SWNT were characterised. In addition, the concept of the pyrolysis of electrospun nanofibres and the use of electrospun fibres as templates for the electrodeless coating of various conducting substrates were also introduced. It was shown that one attractive feature of the ICP is the tailorability of electronic properties. The conductivity of the electrospun fibres was also shown to increase significantly with a decrease in the diameter of the fibre. It was demonstrated that ICP nanofibres are excellent candidates for ultra-sensitive sensors because of the extraordinary high surface area of the electroactive nanofibre.

The nanofibre assemblies were also shown to have unusually low dielectric constants compared to the state-of-the-art ultra low K materials. Various combinations of electrospun nanofibres with and without fillers, including PAN, PAN/CNT, PAN/Fe_2O_3 and PAN/GNP, were examined. The extremely low level of dielectric constant, less than 1.5, can be attributed to nanofibrous assemblies consisting of tortuous, interconnected networks of fibres, which are highly porous without sacrificing surface coverage. This interesting dielectric property of the nanofibrous assemblies will play an important role not only for wearable electronics, but also have significant implications for the next generation of ultra high-density integrated circuits.

Combining the advantages of low cost, control of material characteristics and design flexibility, the polymeric nanostructured fibrous assemblies are serious enabling materials for wearable electronics.

2.6 Acknowledgements

The authors gratefully express their appreciation of the funding provided by the Army Research Office through a Multidisciplinary Research Initiative, which enabled their research on the electrospinning of electroactive nanofibres to be initiated. Support from the Pennsylvania Nanotechnology Institute and from NASA enabled us to expand our research into nanotube-based nanocomposite fibrils. The assistance of Kara Ko in the preparation of this manuscript is greatly appreciated.

2.7 References

1. MIT Media Laboratory, http://web.media.mit.edu
2. GaTech, www.wearables.gatech.edu
3. Post R E et al., *IBM System J.*, 2000, **39**, no. 3 & 4.
4. MacDiarmid A G, *Angew. Chem., Int. Ed.*, 2001, **40**.
5. Gorix Clothing Products Website, www.gorix.com
6. Kuhn H H, Child A and Kimbell W, *Synth. Met.*, 1995, **71**.
7. Nabet B, 'When is small good? On unusual electronic properties of nanowires', ECE Department, Philadelphia, PA-19104.
8. Yao Z et al., *Nature*, 1999, **402**, 6759.
9. Hulteen J C and Martin C R, *J. Mater. Chem.*, 1997, **7**(7).
10. Hongu T and Phillips G O, *New Fibers*, 1997.
11. Formhals A, U.S. Patent # 1,975,504, 1934.
12. Doshi J and Reneker D H, *J. Electrostatics*, 1995, **35**, 151.
13. Gibson P W, Schreuder-Gibson H L and Riven D, *AIChE J.*, 1999, **45**, 190.
14. Ko F K, Laurencin C T, Borden M D and Reneker D H, 'The dynamics of cell-fiber architecture interaction', in *Proceedings, Annual Meeting, Biomaterials Research Society*, San Diego, April 1998.
15. Reneker D H and Chun I, *Nanotechnology*, 1996, **7**(3), 216.
16. Baumgarten P K, *J. Colloid Interface Sci.*, 1971, **36**(1).
17. Sachiko S, Gandhi M, Ayutsede J, Micklus M and Ko F, *Polymer*, 2003, **44**, 5721–5727.
18. Hager B L and Berry G C, *J. Polymer Sci., Polym. Phys. Edn*, 1982, **20**, 911.
19. Norris I D, Shaker M M, Ko F K and MacDiarmid A G, *Synth. Met.*, 2000, **114**(2), 109.
20. ASTM Designation: D4496-87-453.
21. Iijima S, *Nature*, 1991, **354**, 56.
22. Xia Y, MacDiarmid A G and Epstein A J, *Macromolecules*, 1994, **27**, 7212.
23. MacDiarmid A G, Min Y, Wiesinger J M, Oh E J, Scherr E M and Epstein A J, *Synth. Met.*, 1993, **55**, 753.
24. El-Aufy A, Preprint submitted to the 226th *American Chemical Society National Meeting*, New York, NY, September 7–13, 2003.
25. Saito R, Dresselhaus G and Dresselhasus M S, *Physical Properties of Carbon Nanotubes*, 1998.
26. Chun I, Reneker D H, Fong H, Fang X, Deitzel J, Tan N B and Kearns K, *J. Adv. Mater.*, 1999, **31**(36).
27. Huang Z, Wang P C, MacDiarmid A G, Xia Y and Whitesides G M, *Langmuir*, 1997, **13**(6480).
28. Gregory R V, Kimbrell W C and Kuhn H H, *Synth. Met.*, 1989, **28**(1–2), C823.
29. Luoh R, Ko F K and Hahn H T, *Proceeding of the 14th International Conference on Composite Materials* (ICCM-14), July 14–18, 2003, San Diego, CA.
30. http://www.mtiinstruments.com/gaging/appnotes-aci_doc.html
31. http://www.ece.gatech.edu/research/labs/vc/packaging/lectures/lecture5.pdf
32. Pai C S, Velaga A N, Lindenberger W S, Lai W Y, Cheung K P, Bauman F H, Chang C P, Liu C T, Liu R, Diodato P W, Colonell J I, Vaidya H, Vitkavage S C, Clemens J T and Tsubokur F, Interconnect Technology Conference, 1998. *Proceedings of the IEEE 1998 International*, June 1998, pp 39–41.
33. Xu Y, Tsai Y, Tu K N, Zhao B, Liu Q, Brongo M, Sheng G and Tung C H, *Appl. Phys. Lett.*, 1999, **75**(6).
34. Chang K M, Yang J and Chen L, *IEEE Electron Device Letters*, 1999, **20**(4).

35. He L and Xu J, *Proceedings of the 6th International Conference on Solid-State and Integrated-Circuit Technology*, 2001, **1**.
36. Gorman B, Orozco-Teran R, Roepsch J, Dong H, Reidy R and Mueller D, *Appl. Phys. Lett.*, 2001, **79**(24).
37. Hanyaloglu B, Aydinli A, Oye M and Aydi E, *Appl. Phys. Lett.*, 1999, **74**(4).
38. http://domino.research.ibm.com/tchjr/journalindex
39. Purushothaman S *et al.*, Electron Devices Meeting, *2001 IEDM Technical Digest*, 2001.
40. Doshi J and Reneker D, *J. Electrostatics*, 1995, **35**.
41. Kazuhiko E and Tatsumi T, *J. Appl. Phys.*, 1995, **78**(2).
42. Yan F, Farouk B and Ko F K, 'Numerical modeling of an electrostatically driven liquid meniscus in the cone-jet mode', *Aerosol Sci.*, 2003, **34**, 99–116.
43. Fridrikh S V, Yu J H, Brenner M P and Rutledge G C, 'Electrostatic production of nanofibers: control of the fiber diameter', *Proceeding of the 6th International Conference on Textile Composites* (TEXCOMP-6), September 11–13, 2002, Philadelphia, PA.
44. Dror Y, Salalha W, Khalfin R L, Cohen Y, Yarin A L and Zussman E, *Langmuir*, 2003, **29**.
45. Spivak S F, Dzenis Y A and Reneker D H, *Mechanics Res. Commun.*, 2000, **27**(1).
46. Storr G J and Behnia M, *Exp. Therm. Fluid Sci.*, 2000, **22**.

3
Electroceramic fibres and composites for intelligent apparel applications

HELEN LAI-WA CHAN,
KUN LI and CHUNG LOONG CHOY
The Hong Kong Polytechnic University, Hong Kong

3.1 Introduction

Piezoelectric ceramics are smart materials commonly used in mechatronic devices and smart systems.[1-5] They can be used to sense changes in pressure and strain in the environment and can generate electrical responses. These electrical signals can be input to feedback systems to stimulate movements in actuators, to trigger alarms or to switch the systems on or off. They are widely used in devices such as accelerometers, microphones and ultrasonic transducers to detect vibrations, acoustic waves and ultrasound. These electroactive ceramics can be fabricated into fibre form using either the viscous suspension spinning process (VSSP) or by the sol–gel process.[6-24] Sol–gel lead zirconate titanate (PZT) fibres have been incorporated into carbon fibre/polymer composites[13-15] to be used as sensors and actuators in smart structures for aerospace applications. To overcome the brittleness of ceramic fibres and to increase the versatility of their applications, ceramic fibres are often incorporated into polymer matrices to form 1-3 composites.[7, 10, 11, 17-24] Figure 3.1 shows a schematic diagram of a 1-3 ceramic fibre/polymer composite consisting of one-dimensionally connected ceramic fibres embedded in a three-dimensionally connected polymer matrix.[19, 25] The advantages of 1-3 composites are that, compared to piezoceramics, they have a lower acoustic impedance (Z_a = density × velocity), which is closer to the acoustic impedance of water and human tissues. Hence, when used in medical ultrasound and underwater acoustics, 1-3 piezoelectric ceramic/polymer composites can enhance the efficiency of energy coupling. In addition, they can decouple the thickness mode from the lateral modes and enable the input energy to propagate in the direction of thickness.[25-46] In this chapter, the fabrication of samarium and manganese doped lead titanate ($Pb_{0.85}Sm_{0.1}Ti_{0.98}Mn_{0.02}O_3$, PSmT) ceramic fibres by a sol–gel process is described. The fabrication, characterisation and theoretical modelling of ceramic fibre/polymer 1-3 composites are given. Possible ways of using these electroceramic fibres and composites in intelligent apparel applications are also suggested.

42 Wearable electronics and photonics

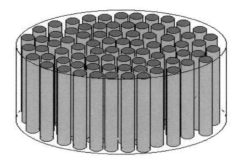

3.1 Schematic diagram of a ceramic fibre/polymer 1-3 composite.[19]

3.2 Fabrication of samarium and manganese doped lead titanate fibres

Lead zirconate titanate (PZT) and lead titanate (PT) are electroceramics widely used in transducer and sensor applications as they can be doped with various dopants to modify their properties. The fabrication of PZT fibres has been described in previous reports.[8, 10, 11, 15, 16, 17, 21, 23] In this work, the fabrication of samarium and manganese doped lead titanate ($Pb_{0.85}Sm_{0.1}Ti_{0.98}Mn_{0.02}O_3$, PSmT) by a sol–gel process[19,20, 22, 24] is described. PSmT is a piezoelectric ceramic with large anisotropy. It has a high thickness electromechanical coupling coefficient k_t and a low planar electromechanical coupling coefficient k_p. Its k_t:k_p ratio is about 10:1 and it has a low relative permittivity (~180). In other words, when a PSmT ceramic disk vibrates in the thickness (z) direction, it has a minimal Poisson's ratio effect

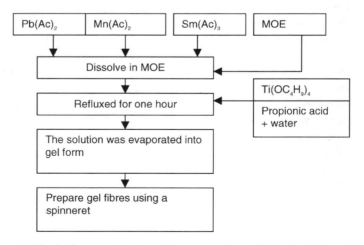

3.2 Block diagram showing the preparation of PSmT gel fibres.[19]

Electroceramic fibres and composites for apparel applications 43

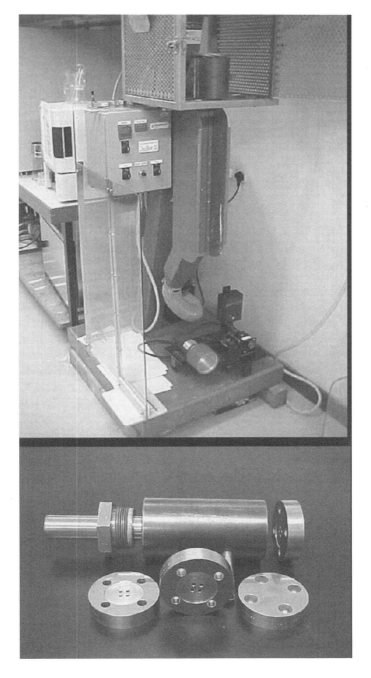

3.3 The fibre spinning machine (OneShot III from Alex James & Assoc., Inc, USA) and the spinnerets.[19]

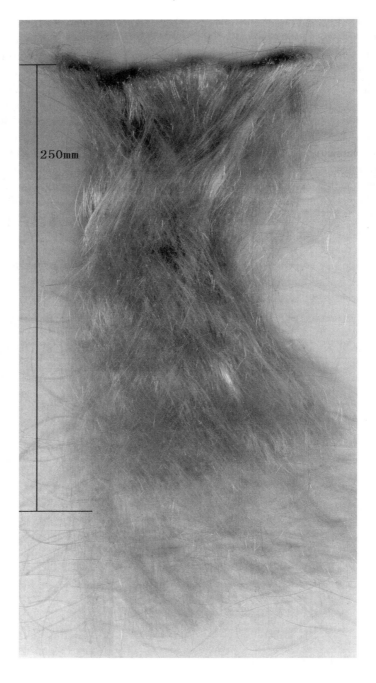

3.4 Photograph of PSmT gel fibres.[19]

and will only induce negligible vibrations in the planar (x and y) directions. Hence, nearly all of the input energy can be delivered in the thickness direction.

In the sol–gel process, as shown schematically in Fig. 3.2, lead acetate trihydrate (with 2% Pb excess), manganese acetate dihydrate and samarium acetate hydrate were heated at 135°C for 8 h to remove the water of hydration, and then dissolved in methoxyethanol (MOE). Titanium n-butoxide was added to the solution and stirred for 30 min to form a samarium and manganese modified lead titanate solution. The solution was then filtered and concentrated. A mixture of water, methoxyethanol and propionic acid (in a molar ratio of 1:5:3) was added. The solution was refluxed for 30 min at 124°C and then evaporated until a gel-like texture had been attained. The gel was poured into the sample chamber of a Model III spinning apparatus manufactured by Alex James and Associates, Greenville, USA. It was extruded into fibre through the die or spinneret (Fig. 3.3) with a 100 μm diameter pinhole (the size of the pinhole can be varied to produce fibres of different diameters) at ~70°C. The fibre was then collected on a spindle. Figure 3.4 shows a bunch of PSmT gel fibres. The gel fibres were tied together, dried and hydrolysed at room temperature for more than a week before being dried at 60–80°C for 48 h. To allow the organic components inside the fibres to escape in a controlled way, the fibres were placed on top of a layer of ceramic powder with the same composition as the fibres. A spoonful of carbon black was scattered around the fibre bundle to reduce the oxygen pressure during heating. The whole setup was then covered with a crucible to pyrolyse at 400°C for 1 h and at 550°C for 1 h. The pyrolysed fibres were calcinated at 850°C for more than 2 h to achieve a complete reaction of the oxides to form PSmT and to densify the fibres.

After calcination at 850°C, the fibres were sintered at 1150°C for 1.5 h. Figure 3.5 (a) to (d) shows the scanning electron micrographs (using a SEM Leica Stereo Scan 440) of the surfaces and the cross sections of a PSmT ceramic fibre. It can be seen that the PSmT fibre is relatively dense without many pores and the average grain size is ~6 μm. The X-ray diffraction (XRD) patterns (Fig. 3.6) measured using an X-ray diffractometer (Philips X'pert PW 3710) show how the crystalline phases of the PSmT ceramics evolved with different heat-treatment temperatures. It can be seen that the PSmT ceramic fibre that was annealed at 750°C for 1 h has started to form a tetragonal structure. When sintered at 1150°C, it has a tetragonal crystal structure with c/a ~1.0439 (where a and c are the axes of the tetragonal unit cell), which is comparable to that of the bulk PSmT ceramics.

3.3   Fabrication of ceramic fibre/epoxy 1-3 composites

A bundle of the sintered ceramic fibres was inserted into a plastic tube with its axis aligned parallel to the tube axis. The tube was then filled with epoxy (e.g. Spurr epoxy, hardness B, Emersion & Cumming, USA). After curing at 70–80°C for 8 h, a ceramic fibre/epoxy 1-3 composite rod was formed. The volume fraction of

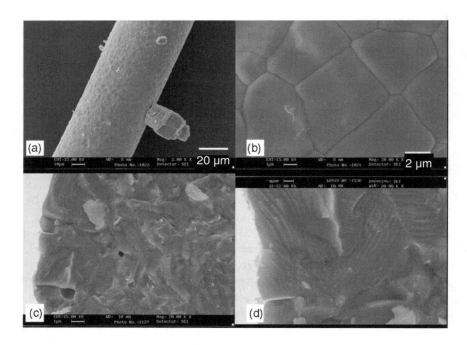

3.5 SEM micrographs of the PSmT ceramic fibres, which were fabricated by hydrolysing gel fibres at room temperature for two weeks and at 60–80°C in moist air for 48 h. The dried fibres were pyrolysed at 400°C for 1 h and at 550°C for 1 h before they were calcinated at 850°C for 2 h and then sintered at 1150°C for 1.5 h. (a) and (b) show the surface; and (c) and (d) the cross sections, of the PSmT ceramic fibre.[19]

ceramics ϕ can be adjusted by changing the number of fibres inserted, and ϕ was calculated by the following equation:

$$\bar{\rho} = \phi\rho + (1-\phi)\overline{\overline{\rho}} \qquad [3.1]$$

where $\bar{\rho}$, ρ and $\overline{\overline{\rho}}$ are the density of the composite, ceramics and epoxy, respectively. The density of the samples was determined by applying the Archimedes principle. The composite rod was sliced into disks in the direction perpendicular to the rod axis. The composite disks were polished and chromium/gold electrodes were deposited on both sides of the disk. To elicit the piezoelectric properties in the fibres, they need to be polarised by applying an electric field to align the dipoles inside the ceramics. The poling was carried out at 110–120°C under an electric field of 4.5 kV mm^{-1} for 15 min. The poled disks were short circuited for more than two days before they were measured using an HP4294A impedance/gain phase analyser and a d_{33} meter. Figure 3.7 shows the SEM micrograph of a PSmT fibre/epoxy 1-3 composite with a 0.4 volume fraction of PSmT. The ceramic fibres in this 1-3 composite have an average diameter of ~ 46 μm. To increase the ceramic

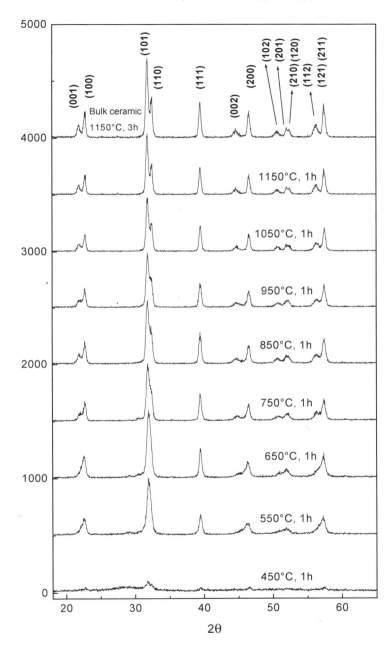

3.6 X-ray diffraction (XRD) patterns of the PSmT bulk ceramics and the PSmT ceramic fibres (crushed into powder before taking XRD patterns) heated at different temperatures.[19]

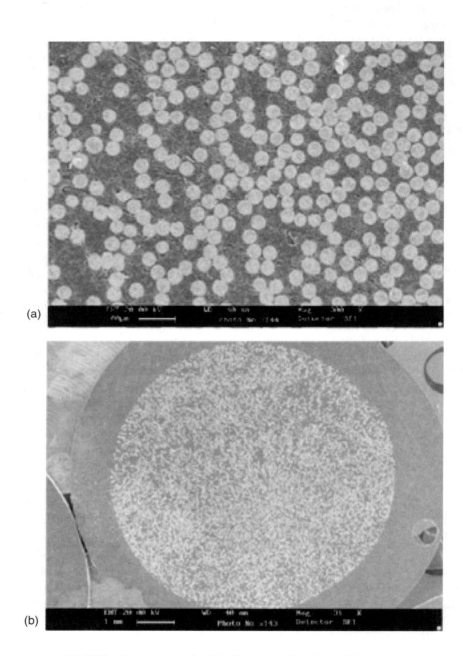

3.7 SEM micrographs of a PSmT ceramic fibre(46 μm)/epoxy 1-3 composite disk. The ceramic volume fraction is 0.4. (a) is an enlarged view of (b).[19]

volume fraction φ in the composite, it was necessary to tie the ceramic fibres tightly together using rubber tape on the outside of the plastic tube. The highest φ attained was 0.68. The volume fraction of ceramics φ can also be measured by integrating the area of the fibres under a scanning electron microscope. From the ratio of the area of the fibre and the epoxy, φ can be estimated.

3.4 Electromechanical properties of ceramic fibre/epoxy 1-3 composites

Figure 3.8 shows the impedance and phase spectra of a PSmT fibre/epoxy 1-3 composite disk with a 4.5 mm diameter containing 35 μm diameter fibres in the frequency range of 10–100 MHz. The volume fraction of PSmT was 0.68. A strong thickness mode resonance was observed and other resonances were very weak. The thickness electromechanical coupling coefficient $\bar{k}_t$ can be evaluated from data obtained from this resonance peak following IEEE standards.[47] It is noted that the impedance at the resonance frequency decreases as the thickness is reduced. In fact, the electrical impedance of the composite can be adjusted to be compatible with that of the driving electronics at the frequency of operation (e.g. adjust to 50 Ω). This is done by changing the ceramic volume fraction, thickness and surface area of the composite. The relative permittivity $\overline{\varepsilon^T_{33}}$, piezoelectric $\overline{d_{33}}$ coefficient, elastic compliance $\overline{s^E_{33}}$ and $\bar{k}_t$ of 1-3 composites with different volume fractions of PSmT fibres were measured. The experimental data are compared with the modelling results in the following section.

3.5 The modified parallel and series model of ceramic/polymer 1-3 composites

In order to use the composite materials in designated applications, it is important to be able to predict their performance. There are a number of proposed models [27, 29–33, 36, 38–40, 45, 46] to evaluate the overall performance of ceramic/polymer 1-3 composites. Here, a modified parallel and series model [29, 30, 36, 39] is described, as this simple model can give reasonable predictions of the materials parameters of 1-3 composites with φ > 0.1. It can be used to calculate the material parameters of PSmT fibre/epoxy 1-3 composites. These figures can then be compared with the experimental data if we assume that the ceramic fibres have properties similar to those of the bulk PSmT ceramics. Alternatively, we can use the measured parameters of the composite and estimate the materials properties of the PSmT fibres by the model calculation.

In this analysis, the material parameters of epoxy are denoted by a double bar on the top of the parameters, while the effective material parameters of the composite are denoted by a bar. The following assumptions have been adopted in the model:[30, 36]

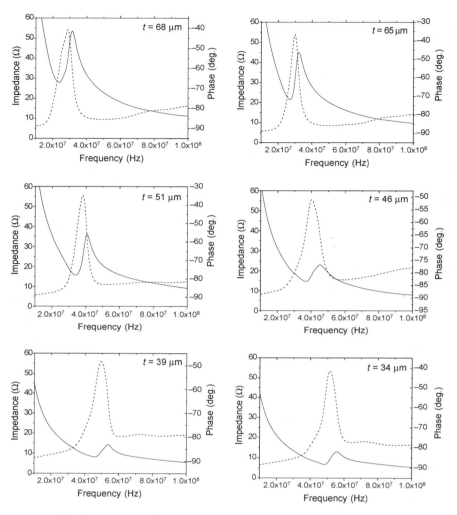

3.8 The electrical impedance and phase spectra of PSmT fibre/epoxy 1-3 composites with different thickness *t*. The 4.5 mm diameter composite disk contains 35 μm diameter fibres and the ceramic volume fraction is 0.68.[19]

(1) The mechanical displacements are parallel to the coordinate axes, so that all shear components vanish in the *x–y* plane:

$$T_4 = T_5 = T_6 = \bar{\bar{T}}_4 = \bar{\bar{T}}_5 = \bar{\bar{T}}_6 = 0 \quad [3.2]$$

$$S_4 = S_5 = S_6 = \bar{\bar{S}}_4 = \bar{\bar{S}}_5 = \bar{\bar{S}}_6 = 0 \quad [3.3]$$

where *T* and *S* denote stress and strain, respectively.

Electroceramic fibres and composites for apparel applications

(2) All fringing fields anywhere inside the materials are zero, giving:

$$E_1 = E_2 = \bar{\bar{E}}_1 = \bar{\bar{E}}_2 = 0 \quad [3.4]$$

$$D_1 = D_2 = \bar{\bar{D}}_1 = \bar{\bar{D}}_2 = 0 \quad [3.5]$$

where E and D are the electric field and electric displacement, respectively.

(3) The composite can be treated as an effective homogeneous medium; both the ceramics and epoxy move together in a uniform thickness oscillation and the vertical strains are the same in both phases (isostrain), giving:

$$\bar{S}_3 = S_3 = \bar{\bar{S}}_3 \quad [3.6]$$

This leads to an effective total stress in the z direction, given by averaging over the contributions of the constituent phases:

$$\bar{T}_3 = \phi T_3 + (1 - \phi)\bar{\bar{T}}_3 \quad [3.7]$$

(4) Electric fields are the same in both the ceramics and epoxy phases in the z direction:

$$\bar{E}_3 = E_3 = \bar{\bar{E}}_3 \quad [3.8]$$

Thus, the effective total electric displacement is the average over the contributions of the two phases:

$$\bar{D}_3 = \phi D_3 + (1 - \phi)\bar{\bar{D}}_3 \quad [3.9]$$

(5) The composite as a whole is laterally clamped, but not for the individual elements:

$$\bar{T}_1 = T_1 = \bar{\bar{T}}_1 \quad [3.10]$$

$$\bar{T}_2 = T_2 = \bar{\bar{T}}_2 \quad [3.11]$$

$$\bar{S}_1 = \phi S_1 + (1 - \phi)\bar{\bar{S}}_1 \quad [3.12]$$

$$\bar{S}_2 = \phi S_2 + (1 - \phi)\bar{\bar{S}}_2 \quad [3.13]$$

The 1-3 composites are considered as isotropic in the x–y plane so that all lateral components are equal $T_1 = T_2$, $\bar{T}_1 = \bar{T}_2$, $S_1 = S_2$, $\bar{S}_1 = \bar{S}_2$. Some important parameters of the composites, e.g. the relative permittivity $\overline{\varepsilon^T_{33}}$, piezoelectric $\overline{d_{33}}$ coefficient and compliance $\overline{s^E_{33}}$, can be derived [19,30,36,39] as:

$$\overline{\varepsilon^T_{33}} = \phi \varepsilon^T_{33} + (1-\phi)\overline{\varepsilon^T_{11}} - \frac{\phi(1-\phi)d^2_{33}}{S(\phi)} \quad [3.14]$$

$$\overline{d_{33}} = \frac{\phi d_{33}\overline{\overline{s_{11}}}}{S(\phi)} \qquad [3.15]$$

$$\overline{s^E_{33}} = \frac{s^E_{33}\overline{\overline{s_{11}}}}{S(\phi)} \qquad [3.16]$$

where

$$S(\phi) = \phi \overline{\overline{s_{11}}} + (1-\phi)s^E_{33} \qquad [3.17]$$

where s^E_{33} is the elastic compliance of the ceramic fibre. Definitions of all of the symbols of piezoelectric ceramics can be found in the IEEE standards for piezoelectricity.[47]

The measured values of $\overline{\varepsilon^T_{33}}$, $\overline{d_{33}}$ and $\overline{s^E_{33}}$ of the PSmT 1-3 composites are plotted, together with the model calculations in Figs 3.9–3.11. Reasonable agreements are obtained showing that this simple model is useful in predicting the materials parameters of 1-3 composites. The measured electromechanical coupling coefficient $\overline{k_t}$ of the composites is found to be higher than that of bulk ceramics, as shown in Fig. 3.12. This is because the ceramic fibres are actually not rigidly clamped by the soft polymer matrix and they can vibrate quite freely. Hence, $\overline{k_t}$ of the composite is close to the k_{33} of a ceramic rod that has a higher value.[30, 36]

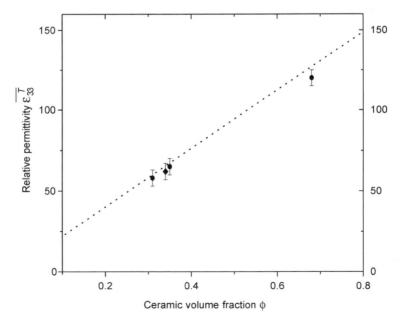

3.9 The relative permittivity $\overline{\varepsilon^T_{33}}$ of the PSmT fibre/epoxy 1-3 composite as a function of ceramic volume fraction ϕ. The symbols and the line represent the experimental data and model calculation from Equation [3.14], respectively.[19]

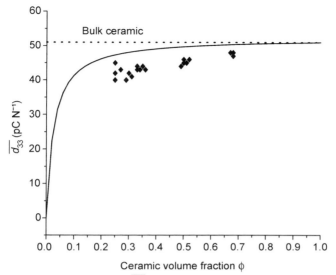

3.10 The piezoelectric $\overline{d_{33}}$ coefficient of the PSmT fibre/epoxy 1-3 composite as a function of ceramic volume fraction ϕ. The symbols and the solid line represent the experimental data and model calculation from Equation [3.15], respectively.[19]

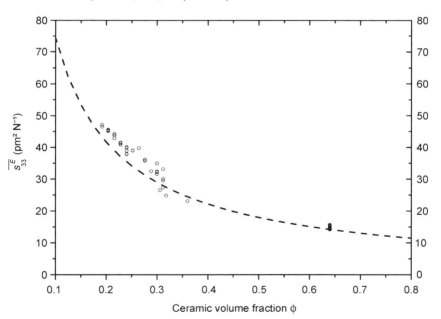

3.11 The elastic compliance $\overline{s^E_{33}}$ of the PSmT fibre/epoxy 1-3 composite as a function of ceramic volume fraction ϕ. The symbols and the line represent the experimental data and model calculation from Equation [3.16], respectively.[19]

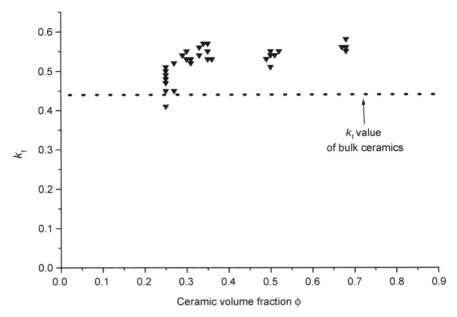

3.12 The measured thickness electromechanical coupling coefficient $\overline{k_t}$ of the PSmT fibre/epoxy 1-3 composite as a function of ceramic volume fraction ϕ. The line represents the k_t of the bulk PSmT ceramics.[19]

3.6 Possible uses of ceramic fibres and composites in intelligent apparel applications

The above discussions show, using PSmT ceramics as an example, the fabrication of ceramic fibres by a sol–gel process and the properties of 1-3 ceramic fibre/polymer composites. Different piezoelectric ceramics and polymers can be selected to optimise the composite performance for the designated applications. These ceramic fibre/polymer 1-3 composites are quite flexible and can be used as acoustic sensors to be incorporated into fabrics and clothing. They can be used to sense vibration, audible sound and ultrasound. They have adjustable density and acoustic impedance, which gives a better match with human tissue. Hence, these acoustic sensors can be used to optimise the transfer of acoustic energy from the human skin to the sensor. This is the distinct advantage of these composites compared to conventional ceramics. For example, they can be fabricated into acoustic sensors attached to a scarf or a belt and can be worn around the neck or wrist for health monitoring. These sensors can pick up heartbeats and breathing rates and, if equipped with wireless communication transmitter, can send the relevant information to a remote receiver. Different sizes of ceramic fibres can also be selected in the fabrication of composites. It is also possible to distribute the fibres in groups to form necessary array patterns. These acoustic arrays can be used to detect sound, and monitor and locate the source of the sound. As 1-3 composites have higher d_{33} coefficients compared with piezoelectric polymer vinylidene fluoride–trifluoroethylene [P(VDF-TrFE)]. They can replace P(VDF-TrFE) copolymer in a number of applications, for example, as actuators and for recovering some of the power in the process of walking in piezoelectric shoe inserts.[48–50] Although 1-3 composites have been widely used in medical ultrasound and underwater acoustics, their use in intelligent apparel and wearable electronics has just begun and more innovative applications can be envisaged in the future.

3.7 Acknowledgements

The support provided by the Hong Kong Research Grants Council and the Centre for Smart Materials of The Hong Kong Polytechnic University is gratefully acknowledged.

3.8 References

1. Berlincourt D A, Curran D R and Jaffe H, 'Piezoelectric and piezomagnetic materials and their function in transducers', *Physical Acoustics*, Vol. 1, Part A, New York, Academic Press, 1964, 169–270.
2. Jaffe B, Cook W R Jr and Jaffe H, 'The piezoelectric effect in ceramics', in J P Roberts and P Popper (eds), *Piezoelectric Ceramics,* London and New York, Academic Press, 1971, 7–23.
3. Mitsui T, Tatsuzaki I and Nakamura E, *An Introduction to the Physics of Ferroelectrics*, London, Gordon and Breach Science, 1976, 1–7.
4. Ikeda T, *Fundamentals of Piezoelectricity*, Tokyo, Oxford University Press, 1990, 119–121.
5. Xu Y H, *Perovskite-type Ferroelectrics: Part I, Ferroelectrics materials and their applications,* Amsterdam and New York, Elsevier Science, 1991, 101–162.
6. Yoshikawa S, Selvaraj U, Brooks K and Kurtz S, 'Piezoelectric PZT tubes and fibers for passive vibration damping', *Proceedings 8th IEEE International Symposium on Applications of Ferroelectrics*, USA, IEEE, 1992, 269–272 .
7. Janas V F and Ting S M, 'Fine-scale, large area piezoelectric fiber/polymer composites for transducer applications', *Proceedings 9th IEEE International Symposium on Applications of Ferroelectrics*, USA, IEEE, 1994, **1**, 295–298.
8. Meyer R J, *Fabrication of Perovskite Lead Zirconate Titanate and Barium Strontium Titanate Fibers using Sol–Gel Technology*, MSc Thesis, The Pennsylvania State University, USA, 1995.
9. Meyer R J, Shrout T R and Yoshikawa S, 'Development of ultra-fine scale piezoelectric fibres for use in high frequency 1-3 transducers', *Proceedings 10th IEEE International Symposium on Application of Ferroelectrics*, Switzerland, IEEE, 1996, **2**, 547–550.
10. Jadidian B, Janas V and Safari A, 'Development of fine scale piezoelectric ceramic/polymer composites via incorporation of fine PZT fibers', *Proceedings 10th IEEE International Symposium on Application of Ferroelectrics*, Switzerland, IEEE, 1996, **1**, 31–34.

11. Meyer R J, *High Frequency (15–70 MHz) 1-3 PZT Fiber/Polymer Composites: Fabrication and Characterization*, PhD Thesis, The Pennsylvania State University, USA, 1998.
12. Glaubitt W, Sporn D and Rainer J, 'Sol–gel processing of functional and structural ceramic oxide fibers', *J. Sol–Gel Sci. Technol.*, 1997, **8**, 29–33.
13. Sporn D, Watzka W, Pannkoke K and Schonecker A, 'Smart structures by integrated piezoelectric thin fibers(I): preparation, properties and integration of fibers in the System Pb(Zr,Ti)O_3', *Ferroelectrics*, 1999, **224**, 1–6.
14. Schonecker A, Keitel U, Kreher W, Sporn D, Watzka W and Pannkoke K, 'Smart structures by integrated piezoelectric thin fibers (II): properties of composites and their physical description', *Ferroelectrics*, 1999, **224**, 7–12.
15. Sporn D, Watzka W, Schonecker A and Pannkoke K, 'Smart structure by integrated piezoelectric thin fibers', in *Piezoelectric Materials: Advances in Science, Technology and Applications*, C Galassi (ed), Netherlands, Kluwer Academic, 2000, 87–97.
16. Steinhausen R, Hauke T, Beige H, Watzka W, Lange U, Sporn D, Gebhardt S and Schonecker A, 'Properties of fine scale piezoelectric PZT fibers with different Zr content', *J. Eur. Ceram. Soc.*, 2001, **21**, 1459–1462.
17. Steinhausen R, Hauke T, Seifert W, Beige H, Watzka W, Seifert S, Sporn D, Starke S and Schonecker A, 'Finescaled piezoelectric 1-3 composites: properties and modeling', *J. Eur. Ceram. Soc.*, 1999, **19**, 1289–1293.
18. Watzka W and Seifert S, 'Dielectric and ferroelectric properties of 1-3 composites containing thin PZT-fibers', *Proceedings 10th IEEE International Symposium on Application of Ferroelectrics*, Switzerland, IEEE, 1996, **2**, 569–572.
19. Li K, *Piezoelectric Ceramic Fibre/Polymer 1-3 Composites for Transducer Applications*, PhD Thesis, The Hong Kong Polytechnic University, 2002.
20. Li K, Wang D Y, Lam K H, Chan H L W and Choy C L, 'Samarium and manganese doped lead titanate ceramic fiber/epoxy 1-3 composites for high frequency transducer applications', *Proceeding of PRICM4 2001*, USA, Japan Institute of Metals, 2001, **2**, 1603–1606.
21. Li K, Lam K H, Wang D Y, Chan H L W and Choy C L, 'Preparation of Li, Nb and Mn doped PZT ceramic fibers and ceramic fiber/epoxy 1-3 composites', *Proceeding of PRICM4 2001*, USA, Japan Institute of Metals, 2001, **2**, 1583–1586.
22. Li K, Chan H L W and Choy C L, 'Samarium and manganese doped lead titanate ceramic fiber/epoxy 1-3 composites for high frequency transducer applications', *IEEE Trans. on Ultrasonics, Ferroelectrics and Frequency Control*, 2003, **50**(10), 1371–1376.
23. Li K, Chan H L W and Choy C L, 'Study of zinc and niobium modified lead zirconate titanate fiber/epoxy 1-3 composites', *Jpn. J. Appl. Phys.*, 2002, **41**, 6989–6992.
24. Chan H L W, Li K and Choy C L, 'Piezoelectric ceramic fiber/epoxy 1-3 composites for high-frequency ultrasonic transducer applications', *Mat. Sci. Eng., B*, 2003, **99**, 29–35.
25. Newnham R E, Skinner D P and Cross L E, 'Connectivity and piezoelectric composites', *Mater. Res. Bull.*, 1978, **13**, 525–536.
26. Klicker K A, Bigger J V and Newnham R E, 'Composites of PZT and epoxy for hydrostatic transducer applications', *J. Am. Ceram. Soc.*, 1981, **64**, 5–8.
27. Gururaja T R, Schulze W A, Cross L E, Newnham R E, Auld B A and Wang Y J, 'Piezoelectric composite materials for ultrasonic transducer applications', *IEEE Trans. Sonics and Ultrasonics*, 1985, **SU-32**(4), 481–498.
28. Gururaja T R, Schulze W A, Cross L E and Newnham R E, 'Piezoelectric composite materials for ultrasonic transducer applications. Part II: Evaluation of ultrasonic medical applications', *IEEE Trans. Sonics and Ultrasonics*, 1985, **SU-32**(4), 499–523.

29. Chan H L W, *Piezoelectric Ceramic/Polymer 1-3 Composites for Ultrasonic Transducer Applications*, PhD Thesis, Macquarie University, Australia, 1987.
30. Chan H L W and Unsworth J, 'Simple model for piezoelectric ceramic/polymer 1-3 composites used in ultrasonic transducer applications', *IEEE Trans. Ultrasonics, Ferroelectrics and Frequency Control*, 1989, **36**(4), 434–441.
31. Hayward G and Hossack J A, 'Unidimensional modeling of 1-3 composite transducers'. *J. Acoust. Soc. Am.*, 1990, **88**(2), 599–608.
32. Smith W A, 'Modeling 1-3 composite piezoelectrics: thickness-mode oscillations', *IEEE Trans. on Ultrasonics, Ferroelectrics and Frequency Control*, 1991, **38**, 40–47.
33. Hossack J A and Hayward G, 'Finite element analysis of 1-3 composite transducers', *IEEE Trans. Ultrasonics, Ferroelectric and Frequency Control*, 1991, **38**(6), 618–629.
34. Slayton M H and Setty H S N, 'Single layer piezoelectric-epoxy composite', *Proceedings IEEE Ultrasonics Symposium*, 1991, 90–92.
35. Waller D J, Safari A and Card R J, 'Woven PZT ceramic/polymer composites for transducer application', *Proceedings 7th IEEE International Symposium on Application of Ferroelectrics*, 1991, 82–85.
36. Smith W A, 'Modeling 1-3 composite piezoelectrics: hydrostatic response', *IEEE Trans. on Ultrasonics, Ferroelectrics and Frequency Control*, 1993, **40**(1), 41–49.
37. Browen L J, Genntilman R L, Pham H T, Fiore D F and French K W, 'Injection molded fine-scale piezoelectric composite transducers', *Proceedings IEEE Ultrasonics Symposium*, 1993, **1**, 499–503.
38. Taunaumang H, Guy I L and Chan H L W, 'Electromechanical properties of 1-3 piezoelectric ceramic/piezoelectric polymer composites', *J. Appl. Phys.*, 1994, **76**(1), 484–489.
39. Chan H L W and Guy I L, 'Piezoelectric ceramic/polymer composites for high frequency applications', *Key Eng. Mat.*, 1994, **92–93**, 275–300.
40. Nowicki A, Kycia K, Iien T D, Gaji O and Klciber M, 'Numerical calculations of the acoustic properties of the 1-3 composite transducers for medical applications', *IEEE Ultrasonics Symposium*, 1995, 1041–1044.
41. Janas V F and Safari A, 'Overview of fine-scale piezoelectric ceramic/polymer composite processing', *J. Am. Ceram. Soc.*, 1995, **78**(11), 2945–2955.
42. Pazol B G, Bowen L J, Gentilman R L, Pham H T and Serwatka W J, 'Ultrafine scale piezoelectric composite materials for high frequency ultrasonic imaging arrays', *IEEE Ultrasonic Symposium*, 1995, 1263–1268.
43. Panda R K, Janas V F and Safari A, 'Fabrication and properties of fine 1,3-composites by modified lost mold method', *Proceedings 10th IEEE International Symposium on Application of Ferroelectrics*, 1996, **2**, 551–554.
44. Hayward G and Bennett J, 'Assessing the influence of pillar aspect ratio on the behavior of 1-3 connectivity composite transducers', *IEEE Trans. on Ultrasonics, Ferroelectrics and Frequency Control*, 1996, **43**(1), 98–108.
45. Takeuchi Y, Nozaki R, Hirata Y, Takada H and Smith L S, 'Novel 1-3 piezo-composites using synchrotron radiation lithography and its application for high frequency medical arrays', *Proceedings IEEE Ultrasonics Symposium*, 1997, **2**, 919–922.
46. Certon D, Casula O, Patat F and Royer D, 'Theoretical and experimental investigations of lateral modes in 1-3 piezocomposites', *IEEE Trans. on Ultrasonics, Ferroelectrics and Frequency Control*, 1997, **44**(3), 643–651.
47. IEEE Standard on Piezoelectricity, ANSI/IEEE Std 176, 1987, 227–273.

48. Starner T, 'Human-powered wearing computing', *IBM System J.*, 1996, **35**(3/4), 618–629.
49. Smalser, *Power Transfer of Piezoelectric Generated Energy*, US Patent Office, Pat No 5 703 474, 1997.
50. Lakic N, *Inflatable Boot Liner with Electrical Generator and Heater*, US Patent Office, Pat No 4 845 338, 1989.

4
Electroactive fabrics and wearable man–machine interfaces

DANILO DE ROSSI, FEDERICO CARPI, FEDERICO LORUSSI, ENZO PASQUALE SCILINGO and ALESSANDRO TOGNETTI
University of Pisa, Italy

RITA PARADISO
Smartex s.r.l., Italy

4.1   Introduction

Multifunctional electroactive fibres and fabrics will give to the traditional textile industry a new added value represented by the possibility of making daily life healthier, safer and more comfortable. They will bring technological advances closer to the public through the realisation of easy-to-use interfaces between humans and devices. This can be achieved by combining advanced microfabrication technologies, material science, textile and electronic engineering in the production of smart clothing, realised by innovative and high knowledge-content textiles that integrate sensing, actuating, electronic and power functions.

The fabrication of such multifunctional interactive fabrics represents a potentially important method for promoting progress, sustainable development and competitiveness in several disciplines, including:

- health monitoring: detecting and preventing diseases
- rehabilitation: restoring lost functions
- health assistance: compensating for disabilities to achieve a higher quality of life
- sports medicine: assessing performance to prevent risks and improve training techniques
- telemedicine and teleoperations: supporting health professionals.

Multifunctional interactive fabrics can also be employed in emerging technology markets, such as:

- wearable wireless communication systems

- localisation and tracking of people
- ergonomics: comfort and safety
- virtual reality: simulation for professional training and entertainment.

There have been a few attempts to design and build prototypes of wearable functional devices. Most of them have taken a limited approach, consisting of attaching off-the-shelf electrical components such as microcontrollers, surface mounted light-emitting diodes (LEDs), piezoelectric transducers, and so on, to traditional clothing material, transforming the cloth into a breadboard of sorts. In fabrics containing conductive strands, these may be used to provide power to devices, as well as to facilitate communication between them. More recently, attempts to construct chip packages directly by a textile process have been reported (Post and Orth, 1997). Other research lines progressed towards the possibility of routing electrical power and communication through suspenders made of a fabric with embedded conductive strands (Gorlick, 1999).

Promising recent developments in material processing, device design and system configuration are enabling the scientific and industrial community to concentrate efforts on the realisation of smart textiles. In fact, all components of interactive electromechanical systems (sensors, actuators, electronics and power sources) can be made of polymeric materials, to be woven directly in textile structures (sensing and actuating micro-fibres) or to be printed or sewn onto fabrics (flexible electronics). In particular, intrinsic sensing, actuating, dielectric or conductive properties, elasticity, lightness, flexibility and the relatively low cost of many electroactive polymers make them potentially suitable materials for the realisation of such systems.

Accordingly, Table 4.1 presents a non-exhaustive list of different polymers used at present for sensing applications and their inorganic conventional counterparts. The list is divided into passive sensors (those that directly convert or amplify the input without a power source) and active sensors (those that require an external power source to convert or amplify the input into a usable output). These two classes of materials should be considered as complementing rather than competing with each other; however, polymers still have to be considered as materials that are at a stage of development. Most of these materials are currently not available in fibre form, and much work has to be done to reach this result.

Table 4.2 reports different polymers currently under investigation for actuating applications and their inorganic counterparts largely used in industrial applications. All of these materials are currently not available in forms compatible with textile technology and much effort is necessary before this goal can be reached.

The aim of this chapter is to give a picture of the potential use of organic materials in the realisation of sensing strain fabrics and of actuating systems. In particular, the early stage of implementation and the preliminary testing of fabric-based wearable interfaces will be illustrated with reference to wearable motion

Table 4.1 Polymers for sensor design and conventional inorganic counterparts

Physical effect	Materials	
	Polymers	Inorganics
Passive sensors		
Piezoelectricity	Polyvinylidene fluoride Polyvinylidene fluoride trifluoroethylene Polyhydroxybutyrate Liquid crystalline polymers (flexoelectricity)	Piezoelectric zirconate titanate Zinc oxide Quartz
Pyroelectricity	Polyvinylidene fluoride Ferroelectric superlattices	Triglycine sulfate Lead-based lanthanum-doped zirconate titanate Lithium tantalate
Thermoelectricity (Seebeck effect)	Nitrile-based polymers Polyphthalocyanines	$Cu_{100}/Cu_{57}Ni_{43}$ Lead telluride Bismuth selenide
Photoelectricity	Polyacetylene/n-zinc sulfide Poly(N-vinyl carbazole)+ merocyanine dyes Polyaniline Poly(p-phenylenevinylene) Polythiophene	Silicon Gallium arsenide Indium antimonide
Electrokinetic	Polyelectrolyte gel Porous ionic polymers	Sintered ionic glasses
Magnetostriction	Molecular ferromagnets	Nickel Nickel–iron alloys
Active sensors		
Piezoresistivity	Polyacetylene Pyrolysed polyacrylonitrile Polyacequinones Polyaniline Polypyrrole Polythiophene	Metals Semi-conductors
Thermoresistivity	Poly(p-phenylenevinylene)	Metals Metal oxides Titanate ceramics Semi-conductors
Magnetoresistivity	Polyacetylene Pyrolysed polyvinylacetate	Nickel–iron alloys Nickel–cobalt alloys
Chemoresistivity	Polypyrrole Polythiophene Ionic conducting polymers Charge transfer complexes	Palladium Metal oxides Titanates Zirconia
Photoconductivity	Copper phthalocyanines Polythiophene complexes	Intrinsic and extrinsic (doped) semi-conductors

Table 4.2 Polymers for actuator design and conventional inorganic counterparts

Physical effect	Materials	
	Polymers	Inorganics
Electronic activation		
Piezoelectricity	Polyvinylidene fluoride	Piezoelectric zirconate titanate
	Polyvinylidene fluoride trifluoroethylene	Zinc oxide
	Polyhydroxybutyrate	Quartz
	Liquid crystalline polymers (flexoelectricity)	
Electrostriction	Dielectric elastomers (acrylic or silicone rubbers)	Barium and lead titanate single crystals
	Polyvinylidene fluoride	Polycrystalline $BaTiO_3$
	Polyvinylidene fluoride trifluoroethylene copolymers	
	Polyvinylidene fluoride hexafluoropropylene copolymers	
Electrostatics	Dielectric elastomers	Silicon
Ionic activation		
Electromechano-chemical	Polypyrrole	–
	Polyaniline	
	Polyelectrolyte gels	
	Polymer–metal composites (IPMC)	
	Carbon nanotubes	

capture systems and a functionalised shirt capable of recording several human vital signs. Although the realisation of a wearable kinaesthetic interface is one of our main aims, it can appear somewhat futuristic. Nevertheless, we have focused our efforts on this application and progressed towards preliminary prototypes.

4.2 Sensing fabrics

Different sensing strategies based on piezoresistive materials are under study and development to realise sensors that can be integrated into elastic fabrics.

4.2.1 Materials and fabric preparation

Different fabrication methods have been used to confer piezoresistive properties to garments. The first approach involves coating conventional fabrics with a thin layer of polypyrrole (PPy, a Π-electron conjugated conducting polymer). PPy is a conducting polymer that combines the good properties of elasticity with mechanical and thermal transduction. PPy-coated Lycra/cotton fabrics that work as strain sensors are prepared using the method reported by Della Santa *et al.* (1999).

Another technique is based on the coating of yarns and fabrics with a mixture of rubber and carbon. The treatment is realised by immersing the material in a solution of rubber and microdispersed phases of carbon. After removal of the excess materials, the conductive elements are immobilised in the structure through a treatment at a temperature of 130°C. The mechanical properties of the final product are affected by the speed of the coating process, the viscosity of the solution and the mutual permeability of materials. Sensors based on carbon loaded rubbers (CLR) realised in this way work as strain sensors.

4.2.2 Characterisation of sensors

The PPy and CLR sensors were characterised in terms of quasistatic and dynamic electromechanical transduction (piezoresistive) properties. Thermal and ageing properties of the sensing fabrics were also preliminarily assessed.

Sensor electromechanical characterisation was performed by exerting uniaxial stretching through rigid links connected to a DC motor and by reading the corresponding variation in electrical resistance. The motor was driven and controlled by an encoder connected to a PC. Quasistatic characterisation was executed by applying small increments of uniaxial stretching, while dynamic characterisation was performed with stepwise stretching. In order to estimate the bandwidth of the sensors, sinusoidal mechanical stimulation at increasing frequencies was applied and the frequency components of the response were investigated.

A simple thermal characterisation procedure was performed to determine how temperature influences the piezoresistive properties of the sensors. The electrical resistance of samples at different temperatures was measured by putting them into an electronically controlled thermostatic cell. To evaluate the ageing behaviour of sensors, the time dependence of the electrical resistance of unstrained threads was evaluated by daily measurements with a digital tester, over a period of one month.

4.2.3 Results and data analysis

PPy-coated Lycra/cotton fabrics

The quasistatic characterisation on PPy-coated fabrics indicates an average gauge factor (GF = $(R-R_0)L_0/(R_0(L-L_0))$, where R and L are the sensor resistance and length, respectively, while R_0 and L_0 are their rest values) of about −13 (negative and similar to that shown by nickel). The numerical value of GF was calculated from a linear interpolation of the data (before saturation) reported in Fig. 4.1.

Despite the fact that the high GF value is suitable for strain gauge implementation, two serious problems affect PPy-coated fabric sensors. The first problem resides in the strong variation with time of the sensor resistance. The second

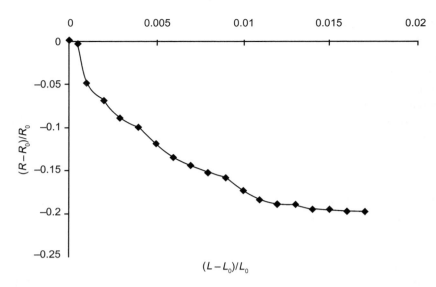

4.1 Typical quasistatic response in terms of percent change in electrical resistance versus uniaxial strain for a PPy-based sensor.

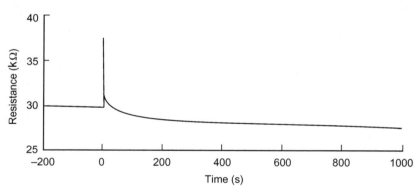

4.2 Response in terms of change in electrical resistance for a PPy-based sensor under a stepwise stretching ($t = 0$s).

problem is the high response time of the sensors; in fact, after the sudden application of a mechanical stimulus the resistance will reach a steady state in a few minutes (see Fig. 4.2); this could restrict the fields of application.

Nevertheless, both limitations have been partially overcome by the following 'ad hoc' coding procedure. Analysing the resistance response in the range of 1 s after the imposition of a stepwise deformation, it is possible to derive the applied strain in an ageing invariant way. We consider a right-angled triangle (Fig. 4.3) where the cathetus height is equal to the excursion of the response peak and the

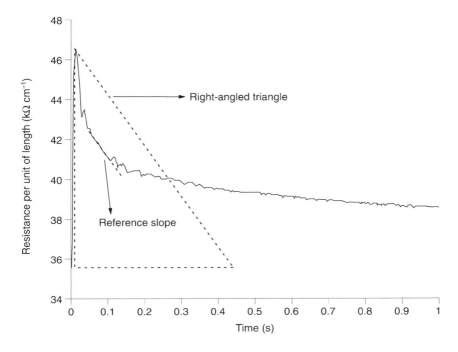

4.3 PPy sensor resistance per unit of length versus time after the application of a stepwise stretching ($t = 0$s). The triangle considered to be useful in data treatment is also traced. The reference slope is given by the slope at the middle point between peak and final value.

slope of the hypotenuse is equal to the time derivative of the resistance, calculated at the middle point between the peak and the final value of the range. It has been demonstrated (Scilingo *et al.*, 2003) that the area of this triangle codifies for the strain independently of the sensor resistance ageing.

PPy-coated fabrics showed a temperature coefficient of resistance (TCR) of about $0.018°C^{-1}$.

Lycra/cotton fabrics coated by carbon-loaded rubber

A GF of about 2.5 was measured for CLR-coated fabrics. The numerical value of GF was calculated from a linear interpolation of the data (before saturation) reported in Fig. 4.4. These values are quite similar to those of metals and are suitable for such sensors to be used in wearable applications.

The behaviour of the fabric sensor when subject to stepwise stimulations at increasing amplitudes was investigated. Different levels of strain were tested (ranging from 1.1 to 13.3%, with incremental variations of 1%). The resistance versus time for three step strains in stretching is reported in Fig. 4.5.

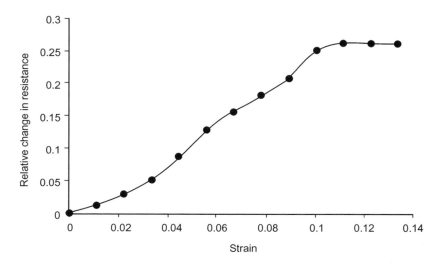

4.4 A typical quasistatic response in terms of percentage change in electrical resistance versus strain for a CLR-based sensor.

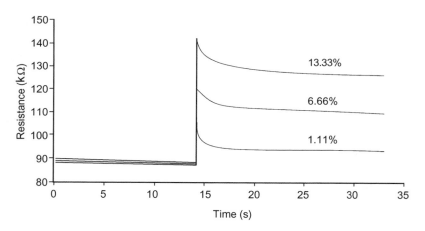

4.5 Resistance response to step strains (1.11, 6.66 and 13.33%) in stretching.

These fabric sensors work well in the range of 1 to 13%. Aside from this range, the response is not univocal and the sensor cannot be used for monitoring kinematic variables. A dual behaviour can be verified with step strains in shortening (see Fig. 4.6).

In this case the fabric was previously stretched to a certain strain, held at that strain for a few seconds in order to attain the steady value of resistance, and then

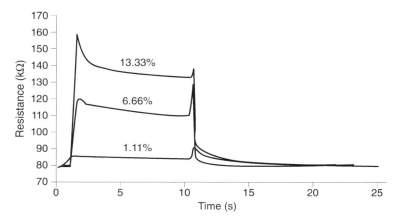

4.6 Resistance response to different step strains (1.11, 6.66 and 13.33%) in shortening.

suddenly relaxed. To establish the sensitivity of the fabric sensor to the imposed strain rates, a set of ramps in stretching as well as in shortening with increasing slope was applied. The results are reported in Figs 4.7 and 4.8. It can be observed that the resistance of the fabric follows the time dependence of the applied strain. Tests on sensor bandwidth showed that once the frequency exceeds 8 Hz, the resistance response is nearly flat. CLR-coated fabrics showed a TCR of about $0.08°C^{-1}$.

4.3 Actuating fabrics

Electroactive polymer actuators are being studied and developed to be embedded into fabrics, to endow them with motorized functions. Three kinds of electroactive materials (dielectric elastomers, conducting polymers and carbon nanotubes) are under investigation in our laboratory.

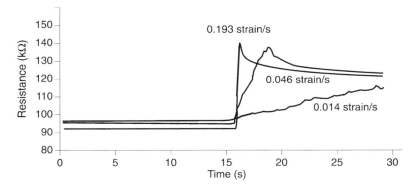

4.7 Resistance response to ramps with increasing slopes in stretching (strain 8.89%).

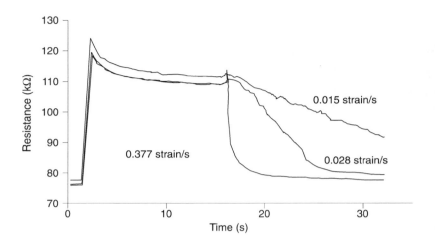

4.8 Resistance response to ramps with different slopes in shortening (strain 8.89%).

4.3.1 Dielectric elastomer wearable actuators

The realisation of actuating devices with fibre geometry implies the need to overcome several difficulties, such as the identification of efficient principles of operation and suitable configurations, the selection of high-performance materials and implementation of custom fabrication processes. Silicone rubbers are being tested as dielectric elastomers to realise high-strain wearable actuators. Dielectric elastomers possess several advantages: linear actuation strains of up to 60%, fast response times (down to tens of milliseconds) and generated stresses of the order of MPa (Pelrine *et al.*, 2000). The price for achieving such performances is represented by the very high driving electric fields (order of 100 V μm^{-1}).

A prototype actuator was realised, made of a silicone hollow cylinder with two compliant electrodes (carbon conductive grease) applied to the internal and external surfaces. Imposing a voltage difference between the electrodes, the polymer sustains an electric field-induced deformation. Preliminary tests showed a longitudinal strain of 3% with a voltage per wall thickness of 85 V μm^{-1} (see Fig. 4.9). The need for high driving electric fields is the actual limitation to the unconditioned use of these actuators.

To improve the performance of actuators made of dielectric elastomers, the possibility of implementing a mechanical amplification of strain is under evaluation. This solution would reduce operation voltages. The system under study is thought to be made of a bundle of cylindrical actuating units, covered by a cylindrical braid mesh (made with flexible but not extensible threads) able to contract when the internal elements impose a radial expansion. The system

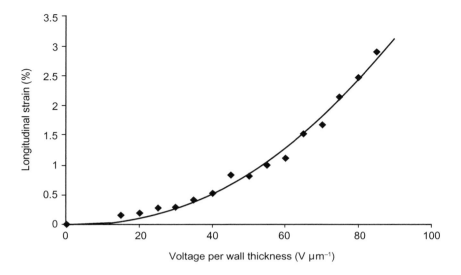

4.9 Longitudinal strain versus applied voltage per wall thickness for a silicone cylindrical actuator.

changes shape, increasing its diameter, decreasing its length and changing the angle α between the axes of the cylinder and the threads.

Two actuating configurations are under study. The first is based on the 'direct' McKibben effect (Chou and Hannaford, 1996). Under electrical stimulation each internal unit undergoes a radial expansion and the mesh produces a strain amplification from the radial direction to the longitudinal one, with tangible shortening and useful contractile forces (see Fig. 4.10).

The amplification factor depends on the mesh structure and its work conditions. It has been shown (De Rossi *et al.*, 2001) that, if the initial value of α is larger than $\pi/4$, the radial expansion is transduced into a linear contraction with an amplification factor larger than 1, as expressed by the relation:

$$\left(\frac{\Delta L_m}{L_m}\right) \approx - tg^2(\alpha) \left(\frac{\Delta R_m}{R_m}\right) \quad \text{for} \quad \left(\frac{\Delta R_m}{R_m}\right) \ll 1,$$

where R_m and L_m are the radius and the length of the mesh, respectively.

The second configuration is based on the 'inverse' McKibben effect. Using radially contracting units and realising a bundle of them with a total diameter higher than the resting mesh diameter, each unit undergoes a radial contraction under electrical stimulation. The mesh produces a strain amplification from the radial direction to the longitudinal one, with lengthening and expanding forces (see Fig. 4.11).

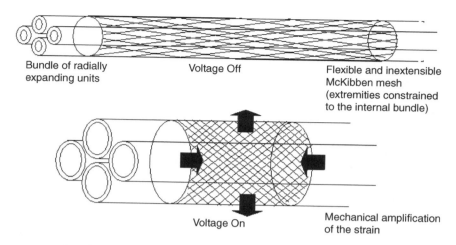

4.10 First actuating configuration: 'direct' McKibben effect.

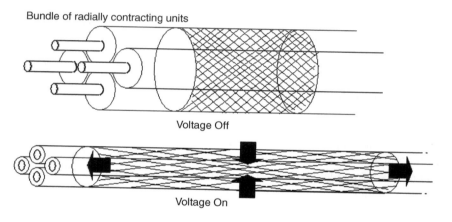

4.11 Second actuating configuration: 'inverse' McKibben effect.

4.3.2 Conducting polymer fibre actuators

Conducting polymer (polyaniline) fibres have been made and tested as actuators (Mazzoldi *et al.*, 2000). They exhibit sizeable active strains (of the order of 1% and more), large active stresses (up to tens of MPa), low driving electrical potential differences (a few Volts) and built-in tunable compliance. Continuum and lumped parameter models for such actuators have been formulated and validated (Mazzoldi *et al.*, 2000), providing a necessary tool for implementing biomimetic control strategies and algorithms. However, at present the use of such actuators is limited by the high value of their response time constants and their short lifetime, both factors being determined by the need for an electrochemical driving force. Attempts are underway to find suitable conductive polymer fibre actuators in order to overcome these limits and to integrate the actuators into active fabrics.

4.3.3 Carbon nanotube fibre actuators

Carbon nanotube fibres have been made and preliminarily characterised as actuators. Carbon nanotubes (the most recent addition to the class of electroactive materials suitable for actuation purposes (Baughman *et al.*, 1999)) are sheets of carbon atoms rolled up into tubes with diameters of around tens of nanometres. Their projected superior mechanical and electrical properties (high actuating stresses, low driving voltages and high energy densities) suggest that superior actuating performances can be expected (Baughman *et al.*, 1999). However, at present the preparation of carbon nanotube fibres has to be much improved to produce fibres able to show all their actuating potential.

4.4 Smart fabrics for health care

An emerging concept of health care, the continuous monitoring of vital signs to provide assistance to patients, is gaining wide acceptance. Wearable non-invasive sensing systems will allow the user to perform everyday activities with minimal training and discomfort.

The development of an integrated sensorised shirt able to record vital signs was described in a previous work (Pacelli *et al.*, 2001). This truly wearable system is conceived to provide continuous remote monitoring of the health status of the patient. The simultaneous recording of vital signs would allow parameter extrapolation and inter-signal elaboration, contributing to the generation of alert messages and synoptic patient tables. Vital signs like electrocardiograms (ECGs) and electromyograms (EMGs) were detected by conductive fabrics, made of steel threads wound round acrylic yarns, while respitrace was acquired (for breathing rate monitoring) by using CLR fabric strips positioned around the trunk, one located at the abdominal level and the other one at the thoracic level. The response of the fabrics was compared with that of commonly used piezoelectric sensors and results were very satisfactory (Fig. 4.12).

4.5 Smart fabrics for motion capture

Truly wearable instrumented garments, capable of recording body kinematic maps with no discomfort for the subject and negligible motion artefacts caused by sensor–body mechanical mismatch, are crucial for several fields of application. The wearable devices described here meet the requirements of comfort and accuracy of motion capture systems. In most applications nowadays the bottleneck is given by devices too cumbersome and invasive for the subject, hence a well-

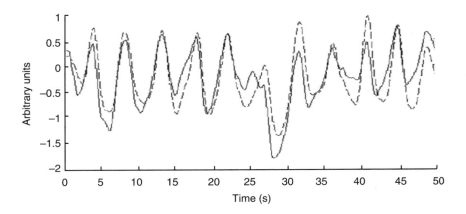

4.12 Respitrace recorded by using the CLR fabric sensor compared with respitrace obtained by a standard piezoelectric sensor (dashed line) currently in use in clinics.

fitting functionalised garment would provide strong advantages. In this context a sensorised leotard and a sensorised glove were fabricated. They are able to permit movements of the arms and upper trunk, and movements of the fingers, respectively. Fields of application range from rehabilitation to sports, from virtual reality to multimedia. The sensing properties of the garment are due to the CLR strain sensors previously described.

4.5.1 Detection of body movements

A few prototypes already realised (De Rossi *et al.*, 2003) have shown a reasonable ability to detect and monitor the position of body segments by reading the mutual angles between the bones. In Fig. 4.13(a) a left flexion of the thorax of a subject wearing a sensorised leotard is shown. Signals detected by sensors placed on the left side (LS) and right side (RS) of the leotard (see Fig. 4.13(a)) are reported in Fig. 4.13(b).

A prototype of a sensorised glove is presented in Fig. 4.14(a) (sensors are visible as black regions on the glove). Each interphalangeal joint is covered by a sensor, while at least two sensors are necessary to detect the position of each metacarpal–phalangeal joint and the carpal–metacarpal joint of the thumb. One degree of freedom was attributed to each interphalangeal joint, two to each metacarpal–phalangeal joint and two degrees of freedom to the trapezium–metacarpal joint. Moreover, relative movements between metacarpal bones have been considered (see the sensor disposition on the glove in Fig. 4.14(a)). The sensors modify their resistance to correspond to the flexions or extensions of each finger. Figure 4.14(b) shows three signals detected by three sensors placed

Electroactive fabrics and wearable man–machine interfaces 73

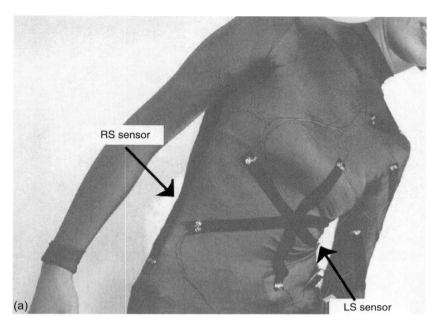

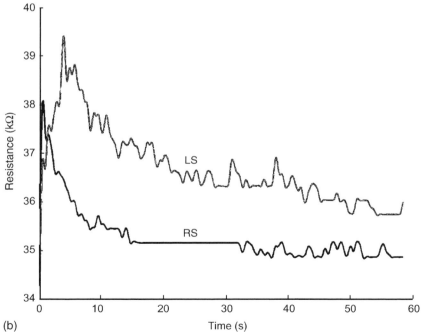

4.13 (a) Left flexion of the thorax of a subject wearing a sensorised leotard and (b) related time change in resistance of the fabric sensors located on the left side (LS) and right side (RS) of the leotard.

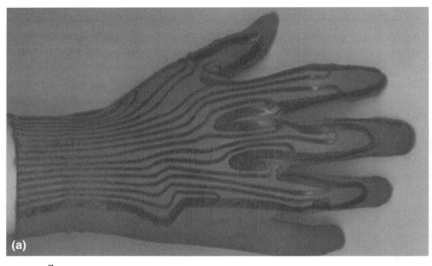
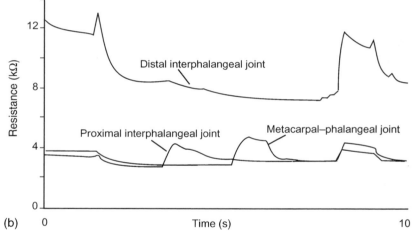

4.14 (a) Sensorised glove. (b) Time changes in resistance of the fabric sensors located on three joints (distal interphalangeal, proximal interphalangeal and metacarpal–phalangeal joint) of the index finger, caused by movements of the finger.

in correspondence to the distal interphalangeal joint, the proximal interphalangeal joint and the metacarpal–phalangeal joint of the index finger.

In the early prototypes sensors had been intuitively located to correspond to each joint in a number equal to the degrees of freedom. In the new generation of prototypes, a large set of sensors is distributed over the garment and a particular strategy of identification and inversion is adopted, as will be discussed in Section 4.6.

4.5.2 Multimedia applications

An application of sensorised fabrics for motion capture is represented by the possibility of triggering music and images by body movements. Strips of sensing fabrics were positioned at joints of the elbow, wrist and shoulder, and a multimedia event was associated with each sensor. The signal gathered from each sensor exhibits an initial peak when it is rapidly stretched. Therefore, its behaviour has to be interpreted and, if necessary, conditioned. As mentioned previously, a preliminary study has confirmed that, when the fabric sensor is submitted to stretching and shortening, its resistance changes. If the mechanical stimulus remains constant, the signal settles at a steady value. This behaviour is suitable for a quantification of the signal. By exploiting all of the dynamic range of variations, a threshold was fixed at the middle point. When this threshold is crossed, a given event starts. Let us focus, for example, on the sensor placed on the elbow. When the arm is completely bent, the sensor is entirely stretched. Corresponding to this position the maximum value of resistance is read. When the arm is stretched, the sensor is at maximum shortening conditions corresponding to the lowest value of resistance. Signals were acquired by a microcontroller-based electronic card via a serial port. During the movement of the arm, images and sounds were switched. The switching frequency depended on both the bandwidth of the sensor signal and the communication speed. Figure 4.15 shows a use of a sensorised glove as a man–computer interface.

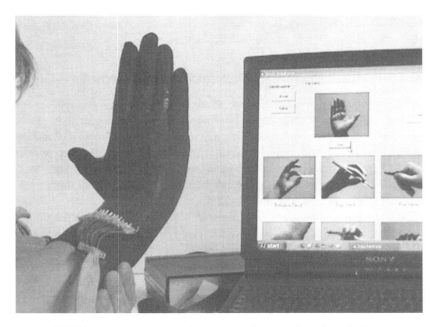

4.15 Use of the sensorised glove as a human interface device.

Another application concerns the world of video games and interfaces (mouse) to control computers. In this case a sensorised glove was fabricated, where each interphalangeal and metacarpal–phalangeal joint was covered by a sensor. Starting from the flat hand position, each sensor modified its resistance in response to a movement of the finger or thumb. To read adduction–abduction of the finger, sensors were placed on the side of the fingers, in correspondence to the interdigital webs. Thumb opposition was detected by reading the output of a sensor crossing the carpal–metacarpal joint on the radial side. For joystick application, four sensors were placed on the index finger, working with one coupled on the first interdigital space and on the opposite side, and the other on the metacarpal– phalangeal volar and dorsal aspect of the index. This geometry was adopted to provide a differential reading of the sensors. Signals were acquired by the joystick port of a PC. This port was designed as an interface with two analogue joysticks. Each joystick had two buttons. The connector of the port enabled the two joysticks to be controlled at the same time. The stick was attached to two 100 kΩ potentiometers. One of the resistors changes its value with a change in the position of the stick along the x axis. The other potentiometer does the same with the y axis. By connecting the fabric sensors to pins relative to the movements along the x and y axes it was possible, through finger movements, to control the mouse pointer on the screen.

A prototype system for interpreting and translating American Sign Language was also developed. Basically, different static configurations of fingers are correlated to specific phrases of language. By means of a database it is possible to establish a small dictionary. In this case sensors are also positioned on the thumb, to allow the recognition of the alphabetical letters.

4.6 Smart textiles as kinaesthetic interfaces

The long-term goal of our research is to develop a family of wearable, bidirectional (sensing and display) man–machine interfaces to be used in surgery and rehabilitation (see Fig. 4.16). To achieve this distant goal, several methodologies and techniques need to be developed in terms of sensing (kinaesthetic), actuation and control. Figure 4.16 refers to a scheme of telerehabilitation, where a bilateral active interface is worn by the patient and is telemetrically controlled and monitored by a medical specialist from a remote position.

Telesurgery is another domain of interest, where an active interface could be worn by the surgeon in a master–slave system to provide better manoeuvrability, dexterity and ergonomic coupling: the surgeon could manoeuvre the system as if he were directly manipulating the remote object itself. Finally, a bilateral interface could be used as a wearable active orthoses for a paralysed arm. In this case, the impaired subject could perform the physical therapy by him- or herself, and the interface could also provide assistance in arm movements.

A reliable kinaesthetic interface should include a system for identifying every

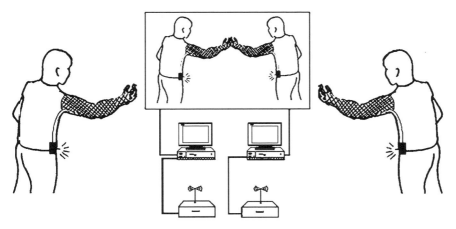

4.16 Scheme of telerehabilitation.

single posture held by the subject wearing the interface itself (De Rossi *et al.*, 2003). With regard to this, a particular strategy of identification and reconstruction ('inversion') was developed, based on a redundant allocation of sensors on the overall interface and an opportune management of the information collected by them, as described below. The adopted technique is, in a certain sense, functional: the final aim is to know which gesture the subject holds and not which individual sensor has modified its status. The technique is based on a calibration phase of the entire system, regardless the single sensor. In fact, when the subject wears the interface for the first time, the information collected by the overall sensor net, following a definite set of known movements, is identified, although the location of the applied deformation is ignored. Data registered during this calibration are then interpolated by a piecewise linear function. From this point of view, redundant sets of sensing fabric patches, linked in different topological networks, can be regarded as a spatially distributed sensing field. Each possible posture and gesture can be reconstructed by simultaneously comparing the present sensing field with the values of the joint variables registered in the calibration phase. As an example, in Fig. 4.17(a) the reconstruction of a target related to the position of an arm is reported.

The abduction–adduction angle and flexion–extension angle of the glenohumeral joint are reported in the abscissa and the ordinate, respectively. Each piecewise line is the solution of the equations holding for significant sensors (responses from sensors which are not influential for the detection of this particular posture degenerate into the entire plane), projected from the entire space of the joint variables onto the plane of the coordinates of the shoulder. The estimated position is given by the intersection of these solutions. In Fig. 4.17(b) the intersection zone is enlarged. Owing to the redundant allocation of sensors, the solution was calculated in the 'least square sense', i.e. by considering the entire

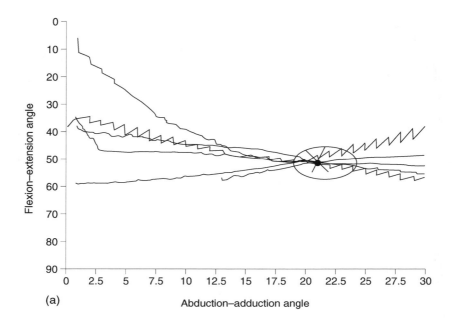

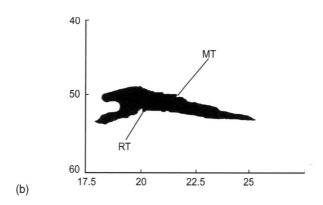

4.17 (a) Traces of the values held by different sensors corresponding to a fixed position. The estimated position is given by the intersection of the traces. (b) High-density zone of probability for the estimated position. The measured target (MT) position differs from the real target position (RT) by less than 4%.

zone with a high density of possible solutions. The position of the target was calculated by the average value of all of the points contained on this zone. The distance between the calculated position of the target and the real one is about 10 cm in a range of about 2.5 m, an error of less than 4%. In other cases larger errors, always less than 8%, were obtained.

4.7 Conclusions

The advanced state of sensing technology can, at present, be considered a realistic scenario for the realisation, in a few years, of truly ready-to-use wearable systems. Much more work has to be done before efficient, reliable and small-size actuators that perform the integration of actuating functions into wearable interactive interfaces can be realised.

4.8 Acknowledgements

The support of DARPA through NICOP (grant no. N000140110280) and of the European Commission through MEGA (FP5 project No. 1999-20410) is gratefully acknowledged.

4.9 References

Baughman R H, Cui C, Zakhidov A A, Iqbal Z, Barisci J N, Spinks G M, Wallace G G, Mazzoldi A, De Rossi D, Rinzler A G, Jaschinski O, Roth S and Kertesz M (1999), 'Carbon nanotube actuators', *Science*, **284**, 1340–1344.

Chou C P and Hannaford B (1996), 'Measurements and modeling of McKibben pneumatic artificial muscles', *IEEE Transactions on Robotics and Automation*, **12**(1), 90–102.

Della Santa A, Mazzoldi A and De Rossi D (1999), 'Dressware: wearable hardware', *Mater. Sci. Eng.*, **C7**, 31–35.

De Rossi D, Lorussi F, Mazzoldi A, Rocchia W and Scilingo E P (2001), 'A strain amplified electroactive polymer actuator for haptic interfaces', in *Smart Structures and Materials 2001: Electroactive Polymer Actuators and Devices*, Y. Bar-Cohen (ed), Proceedings of SPIE, Bellingham, Vol. 4329, 43–53.

De Rossi D, Lorussi F, Mazzoldi A, Orsini P and Scilingo E P (2003), *Active Dressware: Wearable Kinesthetic Systems, Sensors and Sensing in Biology and Engineering*, Springer-Verlag, New York, Chap. 26, 379–392.

Gorlick M (1999), 'Electric suspenders: a fabric power bus and data network for wearable digital devices', *Proceedings of the Third IEEE International Symposium on Wearable Computers*, 18–19 October 1999, 114–121.

Mazzoldi A, Della Santa A and De Rossi D (2000), 'Conducting polymers actuators: properties and modeling', in *Polymers Sensors and Actuators*, Osada Y and De Rossi D (eds), Berlin, Springer, 207–244.

Pacelli M, Paradiso R, Anerdi G, Ceccarini S, Ghignoli M, Lorussi F, Scilingo E P, De Rossi D, Gemignani A and Ghelarducci B (2001), 'Sensing threads and fabrics for monitoring body kinematic and vital signs', *Proceedings of Fibres and Textiles for the Future Conference*, Tampere, Finland, August 2001, 55–63.

Pelrine R, Kornbluh R, Pei Q and Joseph J (2000), 'High-speed electrically actuated elastomers with strain greater than 100%', *Science*, **287**, 836–839.

Post E R and Orth M (1997), 'Smart fabric, or wearable clothing', *Proceedings of the First IEEE International Symposium on Wearable Computers*, 13–14 October 1997, 167–168.

Scilingo E P, Lorussi F, Mazzoldi A and De Rossi D (2003), 'Strain sensing fabrics for wearable kinaesthetic systems', *IEEE Sensors J.*, **3**(4), 460–467.

5
Electromechanical properties of conductive fibres, yarns and fabrics

PU XUE, XIAOMING TAO, MEI-YI LEUNG
and HUI ZHANG
The Hong Kong Polytechnic University, Hong Kong

5.1 Introduction

Textiles now play a crucial role in many engineering applications in addition to their initial use as apparel. Intelligent textiles, representing the new generation of fibres, fabrics and articles, are able to sense changes in their environments, such as mechanical, thermal, chemical, electrical, magnetic and optical changes, and then respond to these changes in predetermined ways.[1-3] Currently, many intelligent materials applied in textiles have been developed, such as photonic fibres, shape memory materials, conductive materials, phase change materials, chromic materials, mechanical responsive materials, intelligent coating/membranes, micro and nanomaterials and piezoelectric materials. With the rapid development of the electrical and, particularly, the electronics industry, flexible electrically conducting and semi-conducting materials, such as conductive polymers, conductive fibres, threads, yarns, coatings and ink, are receiving widespread attention. They are playing a more and more important role in realising lightweight, wireless and wearable interactive electronic textiles.

Highly conductive flexible textiles can be prepared by weaving thin wires of various metals such as brass and aluminium. These textiles have been developed for higher degrees of conductivity. Semi-conductive textiles can be produced in various ways, such as by impregnating textile substrates with conductive carbon or metal powders, patterned printing, and so forth. Conducting polymers, such as polyacetylene (PA), polypyrrole (PPy), polythiophene (PTh) and polyaniline (PAn), offer an interesting alternative. Being chemically or electrochemically doped π-conjugated polymers, they have been found to possess metallic properties.[4] Among them, polypyrrole has been widely investigated owing to its good electrical conductivity, good environmental stability in ambient conditions and because it poses few toxicological problems.[5,6] PPy is formed by the oxidation of pyrrole or substituted pyrrole monomers. Electrical conductivity in PPy involves the movement of positively charged carriers or electrons along polymer chains and the hopping of these carriers between chains. The conductivity of PPy can reach the range 10^2 S cm^{-1}, which is next only to PA and PAn. With inherently versatile molecular structures, PPys are capable of undergoing many interactions.

However, as a conjugated conducting polymer, its brittleness has limited the practical applications of PPy. The processability and mechanical properties of PPy can be improved by incorporating some polymers into PPy.[7,8] However, the incorporation of a sufficient amount of filler generally causes a significant deterioration in the mechanical properties of the conducting polymer, in order to exceed the percolation threshold of conductivity.[9] Another route to overcoming this deficiency is by coating the conducting polymer on flexible textile substrates to obtain a smooth and uniform electrically conductive coating that is relatively stable and can be easily handled.[10,11] Thus, PPy-based composites may overcome the deficiency in the mechanical properties of PPy, without adversely affecting the excellent physical properties of the substrate mterial, such as its mechanical strength and flexibility. The resulting products combine the usefulness of a textile substrate with electrical properties that are similar to metals or semi-conductors.

These products have found wide application in the fields of electromagnetic interference (EMI) shielding, static dissipation, sensing deformation and temperature change, radar-absorbing materials, and so forth.[12-14] They enable the realisation of truly wearable instrumented garments capable of recording surface temperatures and kinetic and dynamic data with no discomfort to the subject and no alteration of signals caused by a mechanical mismatch between the sensor and the body.[15] Currently, some products using interactive electronic textiles have been developed worldwide, such as the Infineon Musical Jacket, Softwitch Electronic Fabrics, Philips Electronic Sportswear Garment, Gorix Electro-Conductive Textile and the Georgia Tech Wearable Motherboard.[16-18] These flexible and ideally conformable products may represent a breakthrough in man–machine interface technology related to virtual reality, teleoperation, telepresence, ergonomics and rehabilitation engineering.[19]

To function as intelligent textiles, the electrical, chemical and mechanical properties of conductive textiles are crucial. Electromechanical behaviour must be thoroughly understood in sensing, actuating and transporting data. In this chapter, we will first briefly introduce conductive textiles, and then focus on the electrical and mechanical properties of PPy-coated conductive fibres/yarn. Based on an understanding of the electrical and mechanical properties of the conductive fibres, the performance of the electrically conductive fabrics will be characterised and evaluated. The relationships between the structural details of the fabric and conductivity will be examined.

5.2 Conductive textiles

Conductive textiles include electrically conductive fibres,[20] fabrics and articles made from them. They can be obtained in various ways.

5.2.1 Metal fibres

Metal fibres can be produced from conductive metals such as ferrous alloys, nickel, stainless steel, titanium, aluminium and copper. Metal fibres are very thin filaments with diameters ranging from 1 to 80 µm. Although highly conductive, metallic fibres are expensive, brittle and heavier than most textile fibres, making it difficult to produce homogeneous blends.

5.2.2 Fibres containing metal, metal oxides and metal salts

Two general methods of coating fibres with conductive metals have been used commercially. One is by chemical plating, the other by dispersing metallic particles at a high concentration in a resin, which is then coated on the surface of the fibre and cured. Semi-conducting metal oxides are often nearly colourless, so their use as conducting elements in fibres has been considered likely to lead to fewer problems with visibility than the use of conducting carbon. The oxide particles can be embedded in surfaces, or incorporated into sheath–core fibres, or react chemically with the material on the surface layer of fibres. Conductive fibres can also be produced by coating fibres with metal salts such as copper sulfide and copper iodide. Metallic coatings produce highly conductive fibres; however adhesion and corrosion resistance can present problems.

5.2.3 Fibres containing conductive carbon

To produce fibres containing conductive carbon, several methods can be used, such as:

- loading the whole fibres with a high concentration of carbon;
- incorporating the carbon into the core of a sheath–core bicomponent fibre;
- incorporating the carbon into one component of a side–side or modified side–side bicomponent fibre;
- suffusing the carbon into the surface of a fibre.

5.2.4 Fibres containing inherently conductive polymers

Based on PAn, PPy and PTh, it is now possible to coat and impregnate conventional fibres with conductive polymers, or to produce fibres from conductive polymers alone or in blends with other polymers.

5.2.5 Conductive yarns and fabrics

Conductive fibres/yarns can be produced in filament or staple lengths and can be spun with traditional non-conductive fibres to create yarns that possess varying degrees of conductivity. Also, conductive yarns can be created by wrapping a non-conductive yarn with metallic copper, silver or gold foil and be used to produce electrically conductive textiles.

Conductive threads are typically finer and stronger than conductive yarns, with controlled conductivity through the placement of stitches. Conductive threads can be sewn to develop intelligent electronic textiles. Through processes such as electrodeless plating, evaporative deposition, sputtering, coating with a conductive polymer, filling or loading fibres and carbonising, a conductive coating can be applied to the surface of fibres, yarns or fabrics. Electrodeless plating produces a uniform conductive coating, but is expensive. Evaporative deposition can produce a wide range of thicknesses of coating for varying levels of conductivity. Sputtering can achieve a uniform coating with good adhesion. Textiles coated with a conductive polymer, such as PAn and PPy, are more conductive than metal and have good adhesion, but are difficult to process using conventional methods.

Adding metals to traditional printing inks creates conductive inks that can be printed onto various substrates to create electrically active patterns. The printed circuits on flexible textiles result in improvements in durability, reliability and circuit speeds and in a reduction in the size of the circuits. The inks withstand bending and laundering without losing conductivity. Currently, digital printing technologies promote the application of conductive inks on textiles.

5.3 Electromechanical properties of PPy-coated conductive fibres/yarns

In this section, we will focus on the electrical and mechanical properties of PPy-coated conductive fibres/yarns, which will be applied in sensing applications under large and repeated deformation, and also establish a basic understanding for modelling the performance of conductive fabrics.

5.3.1 Experimental details

Materials and the preparation of samples

Considering their potential applications in smart textiles, two material systems were examined in the present study: polycaprolactam (PA6) fibres coated with PPy and polyurethane (PU) fibres (Lycra™) coated with PPy. The commercial multifilaments of PA6 with triangular profile were supplied by Dupont, and multifilament PU yarn was supplied by Sunikorn Knitters Limited (Hong Kong). Pyrrole (99%) and ferric chloride hexahydrate ($FeCl_3.6H_2O$) were purchased from the Sigma-Aldrich Chemical Company. All of the chemicals in the highest available grades were used as received without undergoing any purification. The linear density of the PA6 yarn is 702 denier/68F and that of polyurethane yarn is 40 denier/5F.

Electromechanical properties of conductive fibres, yarns and fabrics

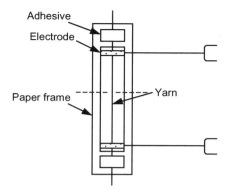

5.1 Schematic drawing of a sample to be tested.

PPy-coated fibres were obtained by the chemical vapour deposition method. The fibres to be coated were fixed parallel each other on a support frame at intervals and soaked in a 50 g L^{-1} FeCl$_3$ solution (used as the oxidising agent) for 10 min. They were then taken out and the residual oxidising solution on the fibre surface was removed by a filter paper. The fibres soaked with oxidising agent were transferred to a glass desiccator in which a beaker containing 10 ml of pyrrole monomer was placed. The desiccator with the fibres and pyrrole was first vacuum-suctioned and then opened to a nitrogen gas atmosphere at room temperature for 24 h. After the vapour deposition process, the PPy-coated fibres were taken out and washed in deionised water for 10 min. They were then put in a desiccator for drying before being measured. Samples of PPy-coated PA6 fibres and PPy-coated PU fibres were both prepared under the same conditions.

Characterisation methods

A scanning electron microscope (SEM) (Lecia Steroscan 440) and a scanning probe microscope (SPM) (API4000/SPA-300HV) were used to observe the surface and cross-section of PPy-coated fibres. The SEM microphotographs were obtained at an accelerating voltage of 5 kV for PPy-coated PU fibres and at a higher accelerating voltage of 20 kV for PPy-coated PA6 fibres. The phase and topographical images were obtained by SPM under ambient conditions.

The electrical resistance of the PPy-coated fibres was measured by the four-probe method with a Keithley 2010 multimeter while the fibres were extended. The load and deformation were obtained and recorded using an Instron mechanical testing system. A single piece of yarn was attached vertically with adhesive (60 mm apart) to a piece of paper with a rectangular hole cut in the centre. Prior to applying a vertical tensile load, the paper was cut horizontally along the dashed line, as shown in Fig. 5.1. The crosshead speed was 5 mm min^{-1} and the gauge length of each specimen was 50 mm. At least five specimens from each sample

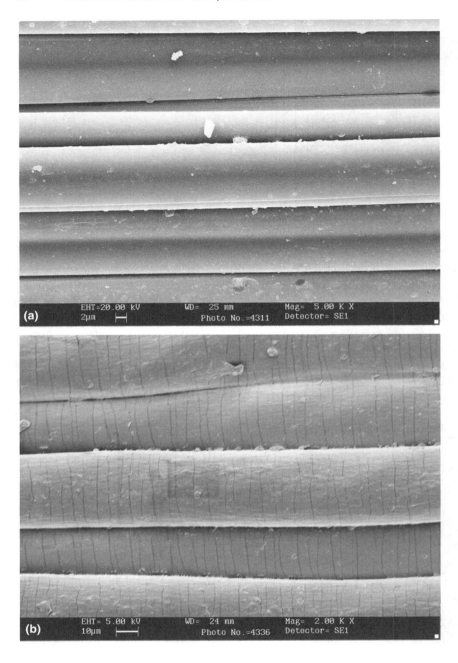

5.2 SEM microphotographs of the (a) PPy-coated PA6 fibres and (b) PPy-coated PU fibres with magnification of 5000 and 2000, respectively.

were tested and the average value was taken. All electromechanical tests were carried out at 20°C and 65%RH (relative humidity).

5.3.2 Experimental results

Microscale observation

From the SEM images, it is believed that the process described for polymerising pyrrole on the surface of PPy-coated fibres encases each single fibre of the textile assembly. For PA6 fibres, the electrically conductive polymer is in a smooth, coherent layer. For PU fibres, some transverse microcracks occurred on the surface, as shown in Fig. 5.2.

The effect of the substrate material on the conductive coating layer was further confirmed by an SPM observation of the cross-sectional view of the fibres, as shown in Fig. 5.3. A continuous PPy conductive domain of 200–300 nm thickness was observed on the surface of the PA6 fibres, whereas there was a thin and discontinuous PPy coating on the PU fibres. On these images, the substrate fibre is shown in the darker region in Fig. 5.3(a) and the lighter region in Fig. 5.3(b). The polymer used to embed the sample is in the lighter region in Fig. 5.3(a) and the darker region in Fig. 5.3(b). The electrically conductive material is in between.

Electromechanical performance under simple tension

By the electromechanical testing system, the relationship between electrical resistance and deformation was obtained. Figure 5.4 depicts typical electrical resistance versus strain curves of two kinds of samples. For the PPy-coated PA6 fibres, it was found that the electrical resistance increased with the increase in the strain, and the relationship between R/R_0 and strain was almost linear under tensile loading until the specimen fractured. However, for the PPy-coated PU fibres, the change in resistance could be divided into two phases. In the initial phase, the resistance increased gradually, followed by a second phase in which the resistance increased non-linearly and rapidly. For the PPy-coated PU fibres, a percolation threshold value existed and it was noticed that the strain threshold was much smaller than the ultimate strain of the PU fibre. As the strain approached the percolation threshold, the current conductive paths became longer and thinner due to transverse microcracks, resulting in the resistance increasing sharply.

As is well known, a sensing material should possess special characteristics, such as linearity, repeatability, sensitivity, and so forth. Among them, linearity between the input and the output signals is one of the most important characteristics. In general, the strain sensitivity of a sensor is represented by the gauge factor, defined as the fraction of the increment in its electrical resistance $\Delta R/R_0$ per unit strain, that is:

$$K \text{ (gauge factor)} = \frac{\Delta R/R_0}{\varepsilon} \qquad [5.1]$$

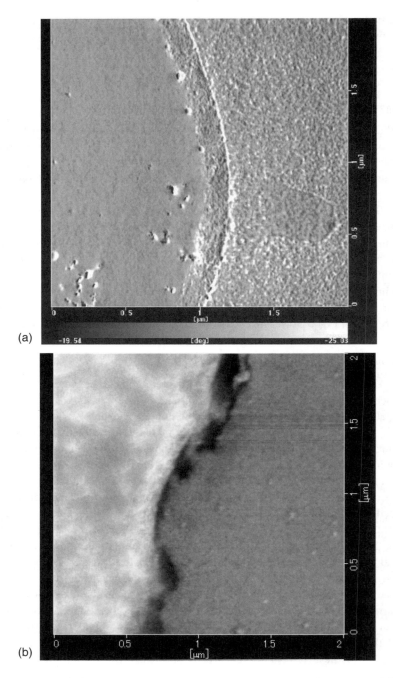

5.3 SPM observation on the cross-sectional view of the fibres, (a) PA6 fibre base; (b) PU fibre base.

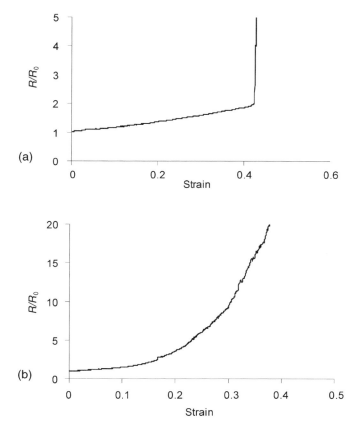

5.4 Typical resistance versus strain curves of (a) PPy-coated PA6 fibres; (b) PPy-coated PU fibres.

where ΔR and R_0 are the change in the resistance and the initial resistance, respectively, and ε is the applied strain. From Fig. 5.5, it was found that the fractional increment in resistance, $\Delta R/R_0$, varies almost linearly with the applied strain, and that the gauge factor of the PPy-coated PA6 fibre is about 2.0 throughout the whole range of strain. Therefore, PPy-coated PA6 fibres possess a good sensing performance. However, as seen from the experimental results shown in Fig. 5.6, PPy-coated PU fibres are not as promising for application as a strain sensor.

The R/R_0 calculated from the change in the dimension of the sample can be calculated from the following equation:

$$R = \rho \frac{L_0}{A_0} \frac{(1+\varepsilon)}{(1-\nu\varepsilon)^2} \qquad [5.2]$$

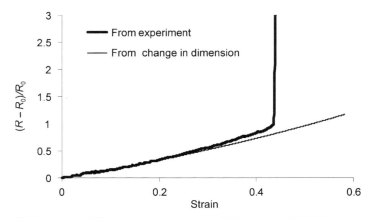

5.5 Typical $\Delta R/R_0$ versus strain curve of PPy-coated PA6 fibres and a comparison of the result calculated from Equation [5.3] and measured from the experiment.

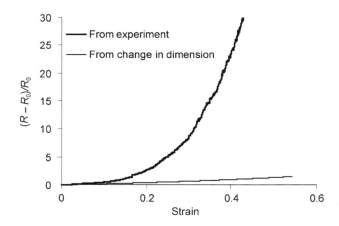

5.6 Comparison of the results of $\Delta R/R_0$ calculated from Equation [5.3] and measured from the experiment for PPy-coated PU fibres.

or

$$R/R_0 = \frac{(1+\varepsilon)}{(1-\nu\varepsilon)^2} \qquad [5.3]$$

where R_0 is the initial electrical resistance equal to $\rho_0 L_0/A_0$, and ν is the Poisson's ratio of PPy-coated fibres when elongated along their longitudinal direction.

Several investigators have specifically studied the mechanical properties of PPy. However, the reports from the literature confirm that the mechanical

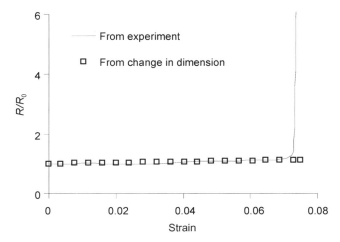

5.7 Comparison of the results of $\Delta R/R_0$ calculated from Equation [5.3] and measured from the experiment for a copper wire.

properties of PPy vary widely from strong, tenacious materials to extremely brittle ones.[4] The composition of the polymer, conditions of polymerisation and the substrate material all have a significant effect on the properties of the polymer. However, owing in part to the intractable nature of PPy film, which makes the characterisation difficult, the relationship is not straightforward. The most commonly reported values are Young's modulus, the tensile strength and the percentage of elongation at break. By taking Poisson's ratio as 0.25 and 0.4 for the PPy-coated PA6 fibre and the PPy-coated PU fibre, respectively, the predictions from Equation [5.3] and measured values of R/R_0 are compared in Figs 5.5 and 5.6. For the PPy-coated PA6 fibres, it was found that the variation in resistance from the change in dimension is slightly smaller than the measured values, but the two curves are quite close to each other when the strain is less than 30%. Therefore, the change in dimension caused by tension is the main cause of the variation in resistance, especially when the applied strain is not large. This behaviour is similar to that of some intrinsically electrically conductive fibres, such as carbon fibres[21] and copper wire, shown in Fig. 5.7. However, the mechanical performance of the PPy-coated fibres is much better than that of these intrinsically electrically conductive fibres. The PPy-coated PA6 fibres provide an excellent sensing performance until the strain of 43%, while intrinsically electrically conductive fibres can only be applied to some situations with small deformation.

In a notable difference from the PPy-coated PA6 fibres, for the PPy-coated PU fibres, the magnitude of $\Delta R/R_0$ caused by the change in dimension only contributes a small percentage of the total variation in resistance and this contribution is further reduced as the applied strain increases. For the PPy-coated PU fibres, the variation in $\Delta R/R_0$ is dominated by the change in its conductivity during tensile deformation.

Performance of conductive fibres under cyclic tension

The sensing function of PPy-coated textiles should be addressed in various environments, such as under cyclic and repeat loads, varying strain rates and temperatures, and so forth. Under cyclic and repeat loading, the electrical resistance may produce a shift, depending on the magnitude of the strain, the number of cycles and the mechanical properties of the carrier employed in a sensor. Figure 5.8 shows typical plots of normalised electrical resistance versus time and applied load versus time for the PPy-coated PA6 fibres, obtained simultaneously during cyclic tension at a maximum strain of 20%, which is equal to 43% of the ultimate strain of the material. From these figures, the applied load was found to have varied consistently as the cyclic number increased. In the first cycle, the electrical resistance increased with loading and decreased with unloading. However, at the end of the first cycle, the electrical resistance could not return to its initial value, R_0. This shift in resistance is attributed to the residual plastic deformation. The shift in electrical resistance is equal to:

$$R_{shift} = \rho \frac{L_p}{A} \qquad [5.4]$$

where L_p is the length of the conductive fibre when the load is unloaded to zero. For the PPy-coated PA6 fibre, L_p was about 55.9 mm. This resistance remained unchanged until the fibre was again extended. Then, reloading caused the electrical resistance to increase, and subsequent unloading behaved in a manner similar to the first unloading. As cyclic loading progressed, both the applied load and the electrical resistance varied repeatly.

Figure 5.9 shows the variations in electrical resistance and applied load as a function of time during the cyclic tension under various unloading strains. At a strain level lower than 30%, the electrical resistance changed periodically along with the applied load, although there was a different residual resistance after the first unloading. The residual electrical resistance was dependent on the magnitude of the unloading strain and nearly did not change with an increase in the number of cycles. However, if the strain reached 35%, equal to 83% of the ultimate strain, the change in resistance became distorted and no longer varied correspondingly to the load.

For the PPy-coated PU fibres, the repeated tension was performed under a 15% strain level. From the experimental observation described earlier on page 87, it is known that many transverse cracks occur on the surface of the conductive fibres under tension. When applying a cyclic load, electrical resistance increased in loading when the cracks opened, and the resistance returned to its original value after unloading when the cracks closed. In the next cycle, the cracks reopened and reclosed at loading and unloading, causing the electrical resistance to vary abruptly, not gradually, as shown in Fig. 5.10.

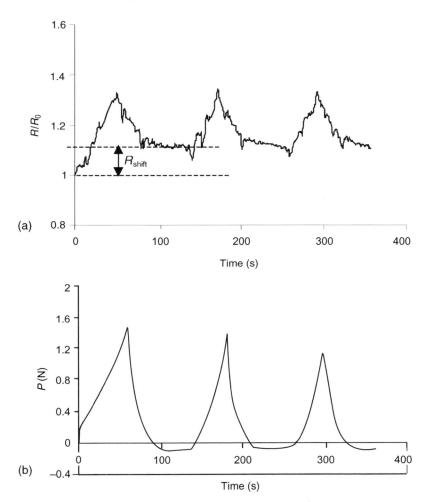

5.8 Variation in (a) electric resistance and (b) applied load during cyclic tension, unloading at strain of 20% (PPy coated PA6 fibres).

Performance of conductive fibres at varying strain rates

Within the capacity of the testing facility, the strain rates examined were 0.17×10^{-3}, 0.17×10^{-2}, 0.17×10^{-1} and 0.17×10^{0} (s^{-1}). A typical group of the load versus strain curves and electrical resistance versus strain curves of the PPy-coated PA6 fibres at different strain rates are given in Fig. 5.11. It is seen that with the increase in the strain rate, the Young's modulus and the strength of the material are enhanced, but the ultimate strain is reduced. Under different strain rates, the change in resistance can be divided into two stages. In the first stage, the relationship between the measured resistance and the applied strain can be

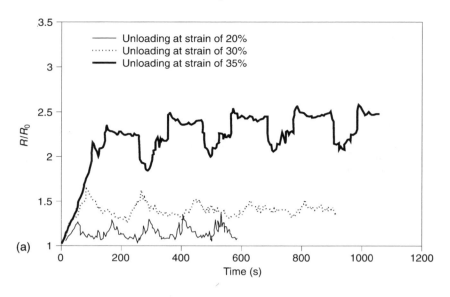

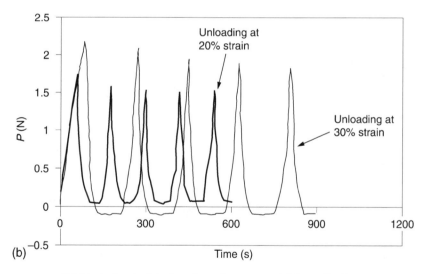

5.9 Variation in (a) electric resistance and (b) applied load during cyclic tension, unloading at different strain levels (PPy-coated PA6 fibres).

reasonably represented by a straight line before a strain threshold, while the threshold appears earlier at higher strain rates. The change in the slope of the straight lines reflects the effect of the strain rate on the performance of conductive fibres. It can be seen that with the increase in the strain rate, the slope of the straight lines increases, while the strain threshold decreases, as shown in Table 5.1.

Electromechanical properties of conductive fibres, yarns and fabrics 95

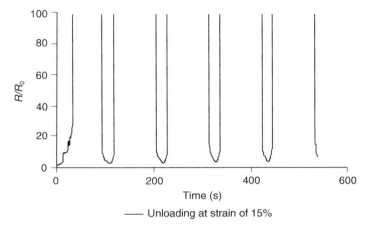

——— Unloading at strain of 15%

5.10 Variation in electric resistance of PPy coated PU fibres during cyclic tension, unloading at a strain level of 15%.

Table 5.1 Slope and threshold under different strain rates

	Strain rate (s^{-1})			
	0.00017	0.0017	0.017	0.17
Slope of the lines	4.03	4.39	5.43	7.98
Strain threshold	0.52	0.47	0.41	0.22

After a linear range, the electrical resistance increases non-linearly until the fibres rupture. This non-linearity is a result of the accumulated microcracks that occurred on the coating layer. The higher the strain rate, the easier it is for accumulated damage to reach a level sufficient to block electrical conductivity of the material.

5.4 Performance of electrically conductive fabrics

The electromechanical properties of a conductive fabric not only depend on the materials used, but on the structure of the fabric. In this section, the performance of two fabric structures, i.e. PPy-coated plain weave fabric and stainless steel knitted fabric, under uniaxial tension will be reported.

5.4.1 PPy-coated woven fabric under unidirectional tension

A plain weave fabric coated with conducting polymer can be generally represented by an electrical network, as illustrated in Fig. 5.12.[22] The resultant total resistance of the plain weave fabric can be simply expressed as:

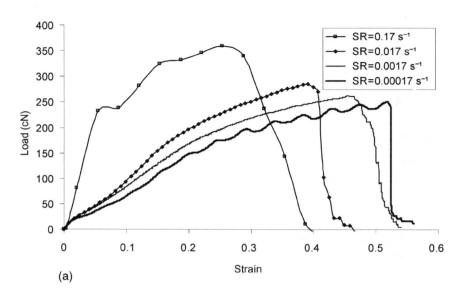

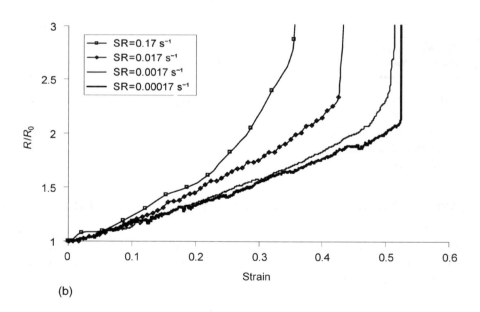

5.11 (a) Load versus strain curves and (b) electrical resistance versus strain curves of PPy-coated PA6 fibres under different strain rates. SR is the strain rate (s^{-1}).

Warp direction: $R_v = \dfrac{\lambda(1 + C_v)(N_p - 1)}{N_e}$ [5.5]

Weft direction: $R_h = \dfrac{\lambda(1 + C_h)(N_e - 1)}{N_p}$ [5.6]

where R_v, R_h are the equivalent resultant resistance of the structure measured in the warp and weft directions, respectively; N_p is the number of picks per length and N_e is the number of ends per width; λ is the resistance of yarn per unit length; C_v and C_h are the crimps of the weave in the warp and weft directions, respectively.

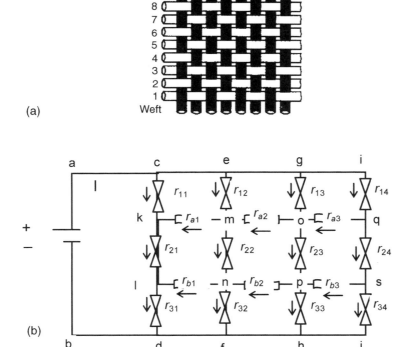

5.12 (a) Plain weave fabric structure. (b) Electrical network representing a segment of the conductive woven fabric.

During the unidirectional tensile deformation (for example in the warp direction), the change in the resistance of the fabric can be expressed as:

$$dR_v = \frac{(N_p - 1)(1 + C_v)}{N_e} d\lambda + \frac{(N_p - 1)\lambda}{N_e} dC_v + \frac{\lambda(1 + C_v)}{N_e} dN_p - \frac{\lambda(N_p - 1)(1 + C_v)}{N_e^2} dN_e \quad [5.7]$$

It can be seen that the change in resistance under large deformation comes from contributions of the yarn and from the geometrical change in the structure of the fabric during tensile deformation, such as change in the crimp of the weave and changes in the densities of the pick and end. Suppose:

$$\lambda = k\varepsilon + \lambda_0 \quad [5.8]$$

then

$$d\lambda = kd\varepsilon \qquad \varepsilon > C_v \quad [5.9]$$

where k is determined from the tensile test of the filaments. Based on the definitions, we have:

$$dC_v = \begin{cases} \dfrac{-(1 + C_v)}{(1 + \varepsilon)^2} d\varepsilon & 0 \leq \varepsilon \leq C_v \\ 0 & \varepsilon > C_v \end{cases} \quad [5.10]$$

$$dN_p = \frac{-N_{po}}{(1 + \varepsilon)^2} d\varepsilon \quad [5.11]$$

$$dN_e = 0 \quad [5.12]$$

where C_v is the weave crimp of the yarn in the fabric and N_{po} is the original number of picks per fabric length. Substituting Equations [5.8]–[5.12] into Equation [5.7], the gauge factor of the electrical conducting fabric in the warp direction can be determined as:

$$\frac{dR_v}{d\varepsilon} = \begin{cases} -\dfrac{1(1 + C_v)}{(1 + \varepsilon)^2} \left[\dfrac{(N_p - 1)}{N_e} + \dfrac{N_p}{N_e} \right] & 0 \leq \varepsilon \leq C_v \\ \dfrac{(1 + C_v)(N_p - 1)k}{N_e} - \dfrac{\lambda(1 + C_v)}{(1 + \varepsilon)^2} \dfrac{N_p}{N_e} & \varepsilon > C_v \end{cases} \quad [5.13]$$

The parameters used in Equation [5.7] can be determined experimentally. For PU yarn and a hand-woven fabric, the parameters used in Equation [5.7] are given in Table 5.2.

From Equation [5.13], it is known that the gauge factor of the conductive woven fabric is inversely proportional to $(1 + \varepsilon)^2$. At a small strain level of the fabric

Electromechanical properties of conductive fibres, yarns and fabrics 99

Table 5.2 Parameters used in Equation [5.7]

k	N_{eo} (ends/inch)	N_{po} (picks/inch)	λ_o (kΩ)
0.43 to 0.47	17	17	8.79

N_{eo} is the original number of ends per fabric width.

(before the filaments were extended), $0 \leq \varepsilon \leq C_v$, the gauge factor is mainly governed by the change in the crimp of the weave as well as by the change in the density of the fabric. The negative sign of the gauge factor indicates that the gauge factor increases with the strain in the decrimping process. Under a large deformation (when the strain is greater than the crimp of the weave), $\varepsilon > C_v$, the gauge factor will be affected by both the variation in the density of the pick and the change in the resistance of the filaments.

5.4.2 Electromechanical properties of stainless steel knitted fabric made from multifilament yarn

An electrically conductive knitted fabric can be depicted as a circuit network, as shown in Fig. 5.13, based on following assumptions:[23]

- filaments are intrinsically conductive and conductivity is a constant that is independent of tensile deformation;
- conductivity at the overlapped points of yarns is related only to the normal force applied;
- the friction of the yarn is negligible;
- two-dimensional hexagon geometry can be used.

The intrinsic resistance of the metallic yarn and the contact resistance both contribute to the electrical properties of the fabric. The intrinsic resistance of the

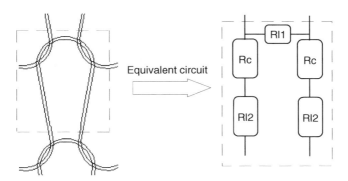

5.13 Circuit network representing a unit loop of the conductive knitted fabric.

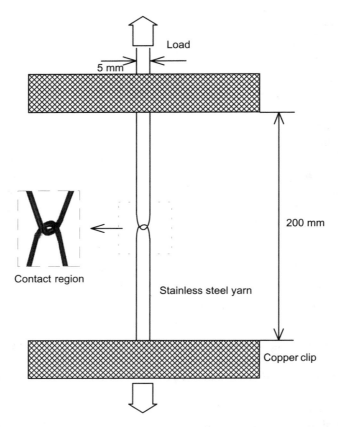

5.14 Experimental setup to measure the contact resistance and contact force at the overlapped point.

metallic yarn can be calculated by Ohm's law and the contact resistance can be determined experimentally. The experimental setup for measuring the contact resistance and contact force is shown in Fig. 5.14. The equivalent resistance of the fabric can then be obtained based on the circuit network.

Based on the experimental results, as shown in Fig. 5.15, it was found that the contacting resistance at the overlapped points decreased with loading. The relationship between the normal force on the two hooked yarns and the contacting resistance can be obtained by the following function:

$$R_c = f(N) \qquad [5.14]$$

where function f was determined experimentally and N is the normal force at the overlapped points.

The equivalent resistance of the fabric can then be obtained based on the circuit

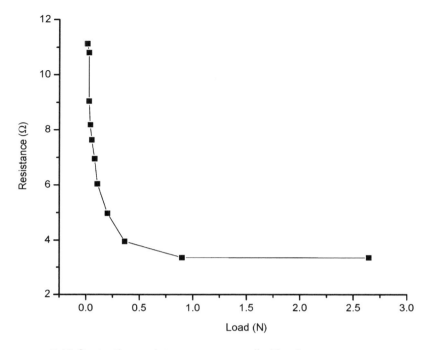

5.15 Contacting resistance versus applied load.

network. The resistances plotted against the load from the experiment and prediction are given in Fig. 5.16. It was found that the contacting resistance at the overlapped points in the knitted fabric governed the change in resistance. The contacting resistance plays a very important role in the sensitivity of the knitted fabric sensor.

5.5 Applications

Electrically conductive textiles make it possible to produce interactive electronic textiles. They can be used for communication, entertainment, health care, safety, homeland security, computation, thermal purposes, protective clothing, wearable electronics and fashion. The details of the applications are listed as follows:[24, 25]

- location/position: infant/toddler/active child monitoring, geriatric monitoring, integrated GPS (global positioning system) monitoring, livestock monitoring, asset tracking, etc.
- infotainment: integrated compact disc players, MP3 players, cell phones and pagers, electronic game panels, digital cameras, and video devices, etc.
- environmental response: colour change, density change, heating change, etc.
- biophysical monitoring (strategic/qualitative assessment only): cardiovascular monitoring, monitoring the vital signs of infants, monitoring clinical trials, etc.

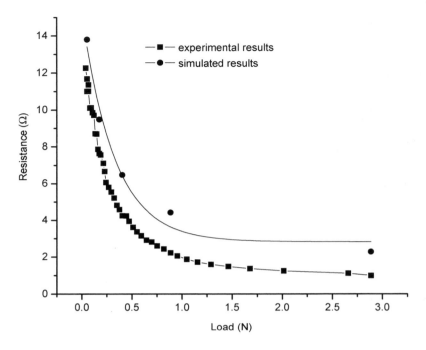

5.16 Comparison of the experimental result and the prediction.

- consumer: health and fitness (professional and amateur; indoor and outdoor activities), home healthcare, fashion, gaming, residential interior design, etc.
- government: military (soldiers, support personnel in the battlefield), space programmes, public safety (fire fighting, law enforcement), etc.
- medical: hospitals, medical centres, assisted-living units, etc.
- commercial: commercial interior design, retail sites, etc.
- industrial: protective textiles, automotive, exposure-indicating textiles, etc.

5.6 Conclusions

This study mainly investigates the electromechanical behaviour of electrically conductive fibres/yarns and fabrics. The following conclusions can be drawn:

- For PPy-coated PA6 fibres, the relationship between the fractional increment in resistance, $\Delta R/R_0$, and the applied strain is reasonably linear within a large strain range, which is of practical importance in sensing applications.
- The variation in resistance for the PPy-coated PA6 fibres results from the change in the dimension of the fibres. By contrast, the variation in resistance with the applied strain for PPy-coated PU fibres is mainly due to damage on the coating layer.
- The strain sensitivity of the conductive woven fabric is inversely proportional

to $(1 + \varepsilon)^2$. Under a low strain level, the gauge factor is mainly governed by the crimp of the weave and density of the fabric. Under a large deformation, it will be affected by both the density of the pick and the change in the resistance of the yarns.
• The contacting resistance in knitted fabric was much larger than the filament resistance itself. The sensitivity of the knitted fabric sensor mainly depends on the contacting resistance and the structure of the fabric.

5.7 Acknowledgement

The financial support extended by the Hong Kong Innovation and Technology Fund is greatly appreciated.

5.8 References

1. Tao X M, *Smart Fibres, Fabrics and Clothing*, Woodhead Publishing, England, 2001.
2. Culshaw B, *Smart Structures and Materials*, Artech House, USA, 1996.
3. Seinivasan A V and Mcfarland D M, *Smart Structures*, Cambridge University Press, UK, 2001.
4. Wallance G G, Spinks M G, Kane-Maguire L A P and Teasdale P R, *Conductive Electroactive Polymers*, CRC Press, New York, 2003.
5. Thiéblemont J C, Brun A, Marty J, Planche M F and Calo P, 'Thermal analysis of polypyrrole oxidation in air', *Polymer*, 1995, **36**, 1605–1610.
6. Omastova M, Pavlinec J, Pionteck J and Simon F, 'Synthesis, electrical properties and stability of polypyrrole-containing conducting polymer composites', *Polym. Int.*, 1997, **43**(2), 109–116.
7. Ruckenstein E and Chen J H, 'Polypyrrole conductive composites prepared by coprecipitation', *Polymer*, 1991, **32**(7), 1230–1235.
8. Truong V-T, Riddell S Z and Muscat R F, 'Polypyrrole based microwave absorbers', *J. Mater. Sci.*, 1998, **33**(20), 4971–4976.
9. Chen Y P, Qian R Y, Li G and Li Y, 'Morphological and mechanical behaviour of an in situ polymerised polypyrrole/Nylon 66 composite film', *Polym. Commun.*, 1991, **32**(6), 189–192.
10. Gregory R V, Kimbrell W C and Huhn H H, 'Electrically conductive non-metallic textile coatings', *J. Coated Fabrics*, 1991, **20**(1), 167–175.
11. Heisey C L, Wightman J P, Pittman E H and Kuhn H H, 'Surface and adhesion properties of polypyrrole-coated textiles', *Textile Res. J.*, 1993, **63**(5), 247–256.
12. Kim M S, Kim H K, Byun S W, Jeong S H, Hong Y K, Joo J S, Song K T, Kim J K, Lee C J and Lee J Y, 'PET fabric/polypyrrole composite with high electrical conductivity for EMI shielding', *Synth. Met.*, 2002, **126**(2–3), 233–239.
13. Xue P, Tao X M, Yu T X, Kwok W Y and Leung M Y, 'Electromechanical behaviour of fibres coated with electrically conductive polymer', *Textile Res. J.*, 2004, **74**(10), 929–936.
14. Hans H and Child A D, 'Electrically conducting textiles', in *Handbook of Conducting Polymers*, 2nd edn, rev. and expanded, 1998, 993–1013.
15. De Rossi D, Della Santa A and Mazzoldi A, 'Dresswear: wearable piezo- and thermo-resistive fabrics for ergonomics and rehabilitation', *Proceedings 19th International Conference-IEEE/EMBS*, Chicago, USA, 1997, 1880–1883.

16. http://www.tut.fi/units/ms/teva/projects/intelligenttextiles/
17. *New Nomads* – An Exploration of Wearable Electronics by Philips, 010 Publisher, Rotterdam, 2000.
18. http://www.softswitch.co.uk/
19. De Rossi D, Della Santa A and Mazzoldi A, 'Dressware: wearable hardware', *Mater. Sci. Eng. C*, 1999, **7**, 31–35.
20. Marchini F, 'Advanced applications of metallized fibres for electrostatic discharge and radiation shielding', *J. Coated Fabrics*, 1991, **20**, 153–166.
21. Cho J W and Choi J S, 'Relationship between electrical resistance and strain of carbon fibres upon loading', *J. Appl. Polym. Sci.*, 2000, **77**, 2082–2087.
22. Leung M Y, Tao X M, Yuen, C W and Kwok W Y, 'Strain sensitivity of polypyrrole-coated fabrics under unidirectional tensile deformation', submitted to *Textile Res. J.*, 2004.
23. Zhang H, Tao X M, and Wang S Y, 'Electro-mechanical properties of stainless steel knitted fabric made from multi-filament yarn under uniaxial extension', submitted to *Textile Res. J.*, 2004.
24. http:// www.vdc-corp.com/
25. Meoli D and May-Plumlee T, 'Interactive electronic textile development: a review of technologies', *J. Textile Apparel, Technol. Management*, 2002, **2**(2), 1–12.

6
Integration of fibre optic sensors and sensing networks into textile structures

MAHMOUD EL-SHERIF
Drexel University, USA

6.1 Introduction

During the twentieth century, tremendous progress was made in the information and telecommunication industries. This progress can be attributed to the development of novel electronics and photonics materials. The development of semi-conductor materials and liquid crystals, processed into microstructural devices, was behind the major achievements in the information and telecommunication industries. A cell phone the size of a matchbox, that receives audio, video and e-mail messages, is an example of such progress.

In the twentyfirst century, the development of electronic and photonic textiles will form the basis for more innovations in electronics and photonics applications, ranging from communication systems and personal computers to biomedical engineering and health-monitoring equipment. Flexible cell phones or televisions with large, flexible screens, which can be folded small enough to fit into a pocket, will be developed in the near future. Health-monitoring shirts for special care will provide direct communication with the patient's physician and will feature feedback control of a through-the-skin drug delivery system. A clip-on textile patch will be used as a cell phone; without using one's hands, one will be able to communicate with anyone, anywhere. Car seats will talk to the passenger to provide comfortable conditions. Uniforms for fire fighters, security guards and special mission personnel will provide all of the information required for their safety and security, as well as transmit remotely information on their health and environmental conditions to a central command facility (El-Sherif, 1997). Novel electronic and photonic systems will be developed that can take advantage of flexible electronic textiles that can conform to the shape of any structure on the ground, in space or under water.

'Smart' textiles are an interesting class of electronics and photonics textiles. They are defined as textiles capable of monitoring their own 'health' conditions and structural behaviour, as well as sensing external environmental conditions and sending the information to other locations. They consist of special types of sensors, signal processors and communication networks embedded into a textile substrate. The conventional sensors and networking systems that are currently available are not technologically mature enough for such applications. New classes of miniature sensors, signal processing devices, and networking links are therefore urgently needed for such applications, and methods for integrating these devices into textile structures need to be developed (El-Sherif et al., 1999).

Responding to the challenges inherent in the development of smart textiles, a highly qualified team was organised in the mid-1990s by the Fibre Optics and Photonics Center of Drexel University and Photonics Laboratories Inc., Philadelphia, Pennsylvania. The team focused on developing a new class of smart textiles composed of integrated fibre optic sensors and networks. Optical fibres are compatible with textile structures and thus present the best choice for the time being. The development of smart textiles requires extensive multidisciplinary research and development (R&D) in textiles and materials engineering, as well as fibre optics and electrical engineering.

A novel case study, based on three major research projects, will be presented in this chapter. These projects focused on the following:

- smart textiles with embedded optical fibres and electric wires for networking and information transmission
- a smart parachute with embedded fibre optic sensors for measuring dynamic structural behaviour during inflation and airdrops
- smart uniforms with embedded fibre optic sensors for detecting environmental conditions and biological threats.

Through these projects, new theories and technologies related to the development of smart textile structures were developed. The R&D work that was conducted ranges from structural design, modelling and analysis to the processing, manufacturing and testing of samples of smart textiles. Specifically, special types of compatible and miniature fibre optic sensors were developed. The methodology to integrate these sensors, as well as optical fibres and fine electrical wires, into various textile-based structures (knitted, woven and non-woven) is under investigation. Integrating lightweight flexible sensors and networks into smart textiles will provide tremendous advantages for many applications. This integration will form the basis for the future development of flexible optoelectronic devices and systems. A number of PhD dissertations and MS theses have been completed, and many technical reports and papers have been published, recording the progress achieved so far.

This chapter is organised into four main sections. The first section presents a short introduction to the design of smart textile structures. To identify the structural design parameters, a new model has been developed for structural analysis and will be discussed in the second section. The third section presents the methodologies that have been developed for integrating optical fibres and electric wires into various textile structures. The last section focuses on the application of smart textiles. The applications are based on the integration of fibre optic sensors and networks into the textile structure to monitor the structural behaviour and environmental conditions. Specifically, two applications are under study. The first involves using a smart parachute canopy to measure the dynamic structural behaviour of parachutes during airdrops and inflation. The second concerns the development of smart uniforms capable of monitoring environmental conditions. These applications require the development of special types of compatible sensors. In the last section, a brief introduction of such sensors, which are strain/stress and chemical sensors, is also presented.

6.2   Smart textiles

The development of smart textiles with embedded optical fibres and electric wires requires a full understanding of the structural behaviour and geometry of textiles. For the proper integration of optical fibres, the processing parameters and final shape conditions of the textile structures need to be carefully identified. Three types of basic textile structures (knitted, woven and non-woven) are under consideration for development as smart textiles.

6.2.1   Knitted structures

There are two main types of knitted structures, warp knitted and weft knitted, as shown in Fig. 6.1. The yarns in a knitted textile material are in the form of interloops that are subjected to very tight bends at a very small radius of curvature. The idea of replacing one of the yarns with optical fibre is not acceptable for these structures. The tight bends may cause extreme losses in optical signals and may even result in their mechanical failure. Therefore, optical fibres used in knitted

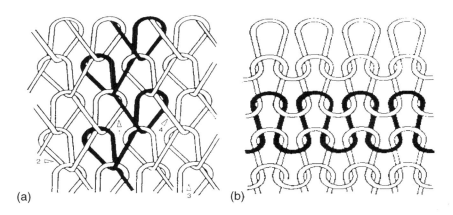

6.1 Knitted textile structure, warp knitted (a) and weft knitted (b).

108   Wearable electronics and photonics

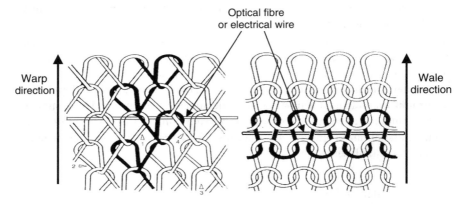

6.2 Optical fibre or electronic wire integrated into the warp (left) or weft (right) textile knitted structures.

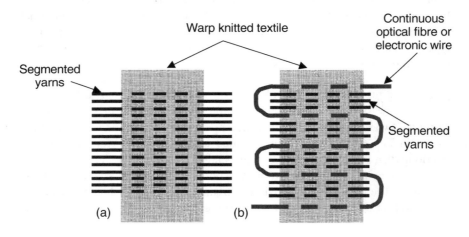

6.3 Schematic diagrams of (a) segmented yarns and (b) continuous optical fibre or wire, taking a serpentine shape.

structures cannot be intertwined in the same way as textile yarns. However, optical fibres and electric wires can be integrated in a straight line, interlacing with the loops. It is possible to integrate optical fibres and wires in a weft or warp knitted textile, as shown in Fig. 6.2. In this case, the integrated optical fibres and wires can have an acceptable path in the textile material without critical bending and mechanical deformation.

To construct a network of optical signals within textile structures, optical fibres and wires must be integrated in a continuous way and not in short segments, as is the case with the yarns shown in Fig. 6.3(a). One of the integration patterns of fibres/wires is in a serpentine shape, as shown in Fig. 6.3(b). Optical fibre can replace some of the segmented yarns at certain locations. The integrated patterns

Integration of fibre optic sensors and sensing networks   109

have to be designed to have a bending curvature of fibres/wires larger than their critical values and also to be suitable for use in knitting machines.

6.2.2 Woven structures

In woven textiles, there are three forms of woven structures: plain, twill and satin. The yarns are interlaced and subjected to bending, as shown in Fig. 6.4. In these woven structures, the yarns are subjected to high bending density. The twill weave, however, has less of a bending curvature than plain or satin structures. Twill yarns, with patterns such as 2/2 or 4/4, could interlace after crossing every two or more strands of transverse yarn. Also, the satin weave has a less dense interlacing structure than the plain weave. The satin weave yarns interlace over or under only one strand of transverse yarn, and then cross two or more strands of transverse yarn. For example, the four-harness weave would interlace over one strand of transverse yarn and then cross four strands of transverse yarn. Therefore, in woven structures, the best integration of optical fibres is in twill weave structures, where the fibre is subjected to the least amount of bending in the curvature. Figure 6.5 presents a cross-sectional view of different weaving structures. Optical fibres (or electric wires) can be integrated in any woven structure in the warp and weft directions, as shown in Fig. 6.6. For example, one of the strands of transverse yarn can be replaced by an optical fibre or electric wire, as shown in Fig. 6.5. However, the best condition for integration is the one with the least bending angles.

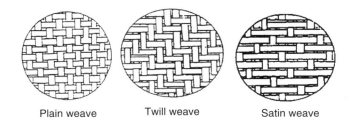

Plain weave   Twill weave   Satin weave

6.4 Three forms of weaving structures.

Plain weave   2/2 Twill weave   4 Harness satin weave

6.5 Cross-sectional view of different weaving structures.

6.6 Optical fibre or wire integrated into a woven fabric in the warp and weft direction.

6.2.3 Non-woven structures

Non-woven textiles are made up of sheet materials composed of more or less randomly oriented segments of fibre bonded together. There are basically two structural features in non-woven structures: the mode of web formation and the mode of bonding. The most common web formation consists of layers of short fibres laid on top of each other. For non-woven structures there are three main types of bonding: entanglement, sticking the fibres together and stitching through the fabric. Thus, the only way for optical fibres and electric wires to be integrated is to place them between sheets (or layers). The orientation of fibres/wires can be in a straight line or with minimum bending curvature. Therefore, the integration of optical fibres in non-woven structures is much easier than in knitted or woven structures.

6.2.4 Studying the machinery

To integrate optical fibres and wires in various textile structures, one must also understand the machinery used in the process. It is very important to evaluate the mechanical motion configuration and process stress parameters exerted on textile yarns during the manufacturing process to achieve the successful integration of optical fibres and wires. Understanding the mechanics involved in the process is essential to selecting the proper way to integrate optical fibres or to modify the process to meet the requirements of optical fibres. Modifications of the process can be achieved by adding an additional track, as a feeder for optical fibres or electric wires. Different types of machines used to manufacture knitted, woven and non-woven fabrics were studied. It was found that some of the mechanical parts, which produce sharp bending in textile fibres during the process, as shown in Fig. 6.7,

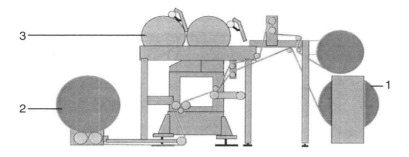

6.7 Schematic diagram of a textile machine, showing the motion of the fabrics/yarns through the major parts (1-2-3), during the manufacturing process.

may need to be modified to limit the angle of the bending of the fibres, as required for optical fibres. Therefore, a study of the machinery may need to be carried out to provide the base of knowledge from which to modify properly the components to adapt the manufacturing process to the requirements of optical fibres.

6.3 Modelling and analysis

Bending is one of the major problems in the integration of optical fibres/wires into textile fabrics, not only during the process but also for the final shape and use of the finished product. Bending may induce mechanical damage and signal loss in optical fibres. Although the bending of optical fibres/wires is a complicated problem, investigations both in theoretical and experimental aspects are very important for the proper design and manufacturing of smart textiles. Therefore, if smart textiles are to be properly designed, the fibre-bending conditions need to be predicted. Unfortunately, the available theories on the design of textile structures are not sufficiently developed to allow predictions of the actual bending angles and curvatures of the yarns at various locations in the textile structures. Optical fibres can be bent within a certain limit without leading to the loss of signals or damage to the fibres. To predict accurately the bending conditions of optical fibres integrated into textile structures, a textile fabric model has been developed. The developed model is designed for a general plainly woven shape; however, it can be modified for application to other textile structures. A plain-weaving unit cell is shown in Fig. 6.8. The diameter of the filling yarns is different from that of the warp yarns. Optical fibre can replace any of these yarns.

This model can be used to predict the geometrical shape and bending of the optical fibres to analyse the effect of the textile's structure on signal processing. In developing this model, relationships of both geometrical and mechanical equilibrium between yarns are considered (Hearle *et al.*, 1969; Olofsson, 1964;

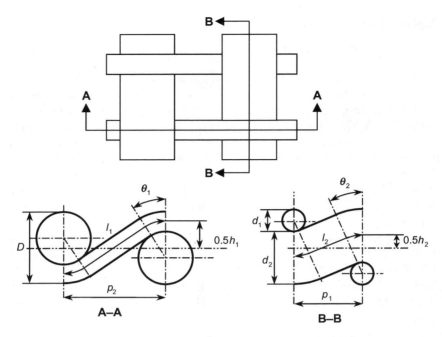

6.8 Orthogonal unit cell showing interlacing geometry for plain weaving structure.

Peirce, 1937). Six equations have been used to govern the orthogonal interlace both geometrically and mechanically. In the calculation to solve the unknown, the Newton–Raphson iteration algorithm (Olofsson, 1964) has been adopted to treat non-linear algebraic equations, and the computer program to perform the calculation has been implemented using C++ computer language. The numerically calculated results have been compared with those from other models.

In this model, circular cross-sectional yarns/fibres, uniform structure and isotropic material properties are assumed. An orthogonal interlace cell consisting of straight and circular portions has been selected to undergo both geometrical and mechanical analyses under the model. The model and the corresponding parameters of the unit cell are shown in Fig. 6.8.

The Peirce geometrical model (Peirce, 1937) gives the following:

$$p_1 = (l_2 - D\theta_2)\cos\theta_2 + D\sin\theta_2 \qquad [6.1]$$

$$p_2 = (l_1 - D\theta_1)\cos\theta_1 + D\sin\theta_1 \qquad [6.2]$$

$$h_1 = (l_1 - D\theta_1)\sin\theta_1 + D(1 - \cos\theta_1) \qquad [6.3]$$

$$h_2 = (l_2 - D\theta_2)\sin\theta_2 + D(1 - \cos\theta_2) \qquad [6.4]$$

$$D = h_1 + h_2 \qquad [6.5]$$

Applying force equilibrium yields:

$$m_1 p_1^2 \sin\theta_1 = m_2 p_2^2 \sin\theta_2 \qquad [6.6]$$

where l_1 is the warp length between two adjacent filling threads (l_1 is not smaller than $D\theta_1$), l_2 is the filling length between two adjacent warp threads (l_2 is not smaller than $D\theta_2$), p_1 is the warp spacing, p_2 is the filling spacing, θ_1 is the weave angle of warp yarn to the plane of the cloth ($0 \leq \theta_1 \leq 0.5\pi$), θ_2 is the weave angle of the filling yarn to the plane of the cloth ($0 \leq \theta_2 \leq 0.5\pi$), h_1 is the displacement of the normal axis of the warp yarn to the plane of the cloth, h_2 is the displacement of the normal axis of the filling yarn to the plane of the cloth, m_1 is the bending rigidity of the warp yarn, m_2 is the bending rigidity of the filling yarn, $D = d_1 + d_2 =$ effective thickness of the fabric, d_1 is the diameter of the warp yarn, and d_2 is the diameter of the filling yarn.

Six independent equations can be used to solve six unknowns. To demonstrate the method, the six unknowns are assumed to be $l_1, l_2, h_1, h_2, \theta_1$ and θ_2. The rest of the parameters are all known values. The Newton–Raphson iteration algorithm has been adopted to treat the non-linear algebraic equations, Equation [6.1] through Equation [6.6]. The equations for solving the problem in matrix form have been derived as follows:

$$\begin{pmatrix} \frac{\partial F_1^i}{\partial l_1^i} & \frac{\partial F_1^i}{\partial \theta_1^i} & \frac{\partial F_1^i}{\partial h_1^i} & \frac{\partial F_1^i}{\partial h_2^i} & \frac{\partial F_1^i}{\partial l_2^i} & \frac{\partial F_1^i}{\partial \theta_2^i} \\ \frac{\partial F_2^i}{\partial l_1^i} & \frac{\partial F_2^i}{\partial \theta_1^i} & \frac{\partial F_2^i}{\partial h_1^i} & \frac{\partial F_2^i}{\partial h_2^i} & \frac{\partial F_2^i}{\partial l_2^i} & \frac{\partial F_2^i}{\partial \theta_2^i} \\ \frac{\partial F_3^i}{\partial l_1^i} & \frac{\partial F_3^i}{\partial \theta_1^i} & \frac{\partial F_3^i}{\partial h_1^i} & \frac{\partial F_3^i}{\partial h_2^i} & \frac{\partial F_3^i}{\partial l_2^i} & \frac{\partial F_3^i}{\partial \theta_2^i} \\ \frac{\partial F_4^i}{\partial l_1^i} & \frac{\partial F_4^i}{\partial \theta_1^i} & \frac{\partial F_4^i}{\partial h_1^i} & \frac{\partial F_4^i}{\partial h_2^i} & \frac{\partial F_4^i}{\partial l_2^i} & \frac{\partial F_4^i}{\partial \theta_2^i} \\ \frac{\partial F_5^i}{\partial l_1^i} & \frac{\partial F_5^i}{\partial \theta_1^i} & \frac{\partial F_5^i}{\partial h_1^i} & \frac{\partial F_5^i}{\partial h_2^i} & \frac{\partial F_5^i}{\partial l_2^i} & \frac{\partial F_5^i}{\partial \theta_2^i} \\ \frac{\partial F_6^i}{\partial l_1^i} & \frac{\partial F_6^i}{\partial \theta_1^i} & \frac{\partial F_6^i}{\partial h_1^i} & \frac{\partial F_6^i}{\partial h_2^i} & \frac{\partial F_6^i}{\partial l_2^i} & \frac{\partial F_6^i}{\partial \theta_2^i} \end{pmatrix} \begin{pmatrix} \Delta l_1^i \\ \Delta \theta_1^i \\ \Delta h_1^i \\ \Delta h_2^i \\ \Delta l_2^i \\ \Delta \theta_2^i \end{pmatrix} = - \begin{pmatrix} F_1^i \\ F_2^i \\ F_3^i \\ F_4^i \\ F_5^i \\ F_6^i \end{pmatrix}$$

[6.7]

where the entries of the right column in Equation [6.7] are given in Equation [6.8]:

$$\begin{cases} -F^i_1 = p_2 - (l^i_1 - D\theta^i_1)\cos\theta^i_1 - D\sin\theta^i_1 \\ -F^i_2 = h^i_1 - (l^i_1 - D\theta^i_1)\sin\theta^i_1 - D(1 - \cos\theta^i_1) \\ -F^i_3 = D - h^i_1 - h^i_2 \\ -F^i_4 = h^i_2 - (l^i_2 - D\theta^i_2)\sin\theta^i_2 - D(1 - \cos\theta^i_2) \\ -F^i_5 = p_1 - (l^i_2 - D\theta^i_2)\cos\theta^i_2 - D\sin\theta^i_2 \\ -F^i_6 = \dfrac{m_1}{m_1}\left(\dfrac{p_1}{p_1}\right)\sin\theta^i_1 - \sin\theta^i_2 \end{cases} \qquad [6.8]$$

The entries of the matrix in Equation [6.7] are given in Equation [6.9]. Therefore, the terms in Equation [6.8] are substituted, and the matrix given by Equation [6.7] becomes Equation [6.9]:

$$\begin{pmatrix} \cos\theta^i_1 & (D\theta^i_1 - l^i_1)\sin\theta^i_1 & 0 & 0 & 0 & 0 \\ \sin\theta^i_1 & (l^i_1 - D\theta^i_1)\cos\theta^i_1 & -1 & 0 & 0 & 0 \\ 0 & 0 & -1 & -1 & 0 & 0 \\ 0 & 0 & 0 & -1 & \sin\theta^i_2 & (l^i_2 - D\theta^i_2)\cos\theta^i_2 \\ 0 & 0 & 0 & 0 & \cos\theta^i_2 & (D\theta^i_2 - l^i_2)\sin\theta^i_2 \\ 0 & -\dfrac{m_1}{m_2}\left(\dfrac{p_1}{p_2}\right)^2\cos\theta^i_1 & 0 & 0 & 0 & \cos\theta^i_2 \end{pmatrix} \begin{pmatrix} \Delta l^i_1 \\ \Delta\theta^i_1 \\ \Delta h^i_1 \\ \Delta h^i_2 \\ \Delta l^i_2 \\ \Delta\theta^i_2 \end{pmatrix} = \begin{pmatrix} F^i_1 \\ F^i_2 \\ F^i_3 \\ F^i_4 \\ F^i_5 \\ F^i_6 \end{pmatrix}$$

[6.9]

The solution for Equation [6.9] is obtained through iteration, where the superscript i is the iteration counter, such as in $l^{i+1}_1 = l^i_1 + \Delta l^i_1$. The convergence of the solution during the process of iteration is measured by the fraction of the Euclidean norm, and the iteration will be terminated when it satisfies the required tolerance (*Tor*). Note that any non-zero estimate for the unknowns can be used as an initial estimate.

To demonstrate the validity of the model, the numerical calculations were performed using self-developed software (FABCAL v2.0). The result based on the current model was compared with the result from either the Peirce model (Peirce, 1937) or the Olofsson model (Olofsson, 1964). If the desired tolerance is set to *Tor* = 10^{-8}, it only takes about six or seven iterations to make the solution converge within a second in most current personal computers. After the solution is obtained, the warp crimp c_1 and filling crimp c_2 can be readily calculated using the following equations:

$$c_1 = \frac{l_1}{p_2} - 1 \qquad [6.10]$$

$$c_2 = \frac{l_2}{p_1} - 1 \qquad [6.11]$$

In addition, the bending force can be calculated as:

$$S_1 = S_2 = \frac{\pi E_1 d_1^4 \sin\theta_1}{8p_2^2} = \frac{\pi E_2 d_2^4 \sin\theta_2}{8p_1^2} \qquad [6.12]$$

where E_1 = Young's modulus of the warp yarn and E_2 = Young's modulus of the filling yarn.

In summary, based on this developed model and the numerical analysis performed, the bending angle of embedded optical fibres can accurately be predicted for various woven structures (Zhao *et al.*, in press). As soon as the bending angles are known, the transmission of optical fibres can be evaluated theoretically and experimentally, before the fibres are integrated in textile structures. Also, proper fibre materials can be selected based on geometrical–mechanical system parameters.

6.4 Manufacturing of smart textiles

In developing the methodology to integrate fibres/wires, the proper fibre/wire must first be identified and selected. The optical, mechanical and electrical properties of these fibres/wires must be characterised. One of the major issues in the process of integration is the bending condition of the fibres/wires during the manufacturing process and in the final textile products. Bending with a small curvature on the fibre/wire may result in high optical loss or in increasing the wire resistance and even breaking of the fibre/wire. Thus, critical bending parameters have to be defined for different types of fibres/wires before the fibres/wires are applied in the process of integration. Therefore, techniques for evaluating optical fibres and electric wires have to be developed. In this section, newly developed testing methods are discussed and a number of examples of the integration of fibres/wires in knitted, woven and non-woven textiles are presented.

Different types of warp knitting machines (or weft knitting) have the capability to integrate optical fibres into the filling direction of the fabric during the knitting process. The optical fibres can be integrated at specific intervals of time using the machine's software program. Specific time intervals for feeding can be set for different kinds of integration processes to limit the bending curvature at the end of the lines.

With regard to weaving machines, most such advanced machines are computerised and can be used to produce plain, twill or satin woven structures with different patterns and densities of yarn. The yarn density is expressed as picks per inch (PPI). A large number of samples were manufactured. The yarn densities of the processed samples ranged from 25 to 200 PPI. Regular textile machines were used

6.9 Schematic of optical fibre.

with no modification or with minor modification by adding an extra loom for the integration of optical fibres/wires.

Non-woven fabrics are produced by heat bonding or by stitching the polyester sheets together. In this case, optical fibres can be embedded between the sheets. The whole layered sample is then fed into a heated roller machine. In another process, the bonding of the polyester sheets is achieved by using a binder, while the optical fibre is positioned between the sheets.

6.4.1 Selection and testing of optical fibres and electric wires

Optical fibres

Optical fibres are cylindrical dielectric waveguides, consisting of three layers, in general, the core, cladding and coating/jacket layers, as shown in Fig. 6.9. The core and cladding are both made of dielectric transparent materials. Both ceramic materials, such as silica and alumina, and polymers, such as polystyrene and polymethyl methacrylate (PMMA), can be used as core or cladding materials. The coating/jacket is used to protect and mechanically support the fibres; thus, they are made of non-transparent polymers and may consist of more than one layer.

The optomechanical properties of the fibres are the major characteristics considered in smart textile applications. The selected fibres have to have the properties of flexibility and small bending loss. There are many types of fibres on the market, from single-mode to multimode fibres, and from all-silica to all-polymer fibres, with a variety of combinations of material properties and core and cladding geometry. The values of the fibre in terms of Young's modulus, maximum strength and Poisson's ratio are important parameters for embedded applications.

Different models have been developed to estimate the bending loss in single-mode or multimode fibres. Since bending is a complicated problem, these models are based on certain assumptions and are fairly complex. Therefore, the experimental testing of specific types of fibres is very important in determining the limitations of bending. In general, when the bending radius is less than the critical bending radius, there is a dramatic increase in bending loss. Since no previous work has appeared in the literature explaining how these tests can be conducted or

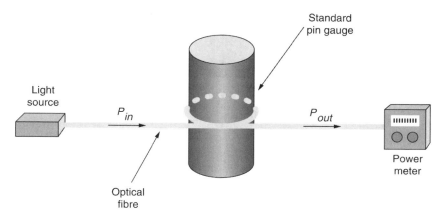

6.10 Experimental setup of bending loss measurement for optical fibres.

discussing any similar applications, part of this study is concerned with ways of developing the testing methods and standard parameters for weaving optical fibres into textile structures.

Different types of (compatible) single-mode and multimode fibres were selected and tested for their optical signal loss at different bending curvatures. An experimental setup was designed for a test of bending loss in fibres, as shown in Fig. 6.10. The optical fibre is wrapped around a standard pin gauge. A light source is connected to the fibre from one end and a power meter is connected to the other end. For each type of fibre, a test was performed at a different bending radius (R) and a different number of turns around standard pin gauges. The diameter of the standard pin gauges used ranged from 25 to 2.5 mm.

Since bending loss in fibres is related to the wavelength of transmitted light, both a 850 nm and a 1550 nm light source were used to test each type of fibre. The test was applied to a large number of fibres to study the behaviour and properties of each under various bending conditions. The optical signal loss induced by wrapping each fibre one cycle around different sizes of standard pin gauges was measured and recorded. Samples of the test results performed on five multimode optical fibres at 850 nm are shown in Fig. 6.11, and the general properties of those fibres are listed in Table 6.1 for comparison. These fibres all have different geometries and different material properties.

The optical loss for a number of cycles (n) at a specific bending diameter (d) is calculated according to the following equation:

$$\alpha(nd) = -10\log \frac{P}{P(nd)}$$

where P and $P(nd)$ are the output powers before and after the fibre is wrapped (n) turns around standard pin gauges of diameter (d), respectively.

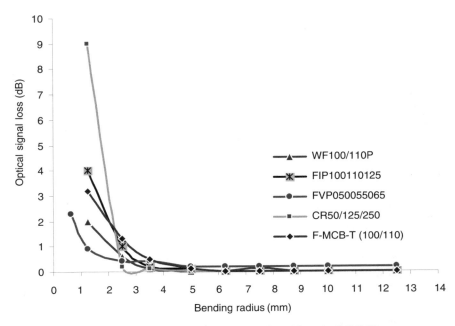

6.11 Optical signal loss of five types of multimode (MM) fibres, measured at different bending curvature using 850 nm light source.

It can be inferred from Fig. 6.11 that the loss is negligible for all selected multimode (MM) fibres when their bending curvature is larger than 5 mm; however, the loss of different types of fibres shows diversity when they are bent smaller than 5 mm. For one fibre, FVP050055065, optical loss starts to increase quickly when the bending radius is smaller than 2 mm. However, mechanical testing shows that this fibre can be bent to 0.625 mm without experiencing mechanical failure or breaking. The critical bending radii (defined as the value when fibre optical bending loss starts to increase) of the other four types of fibres,

Table 6.1 Properties of the multimode optical fibres shown in Fig. 6.11

Fibre part number	Size (core/clad/jacket) (µm)	Materials (core/clad/jacket)	NA	Tensile stress (kpsi)
FVP050055065	50/55/65	Silica/silica/polyimide	0.22 (sp)	100
FIP100110125	100/110/125	Silica/silica/polyimide	0.22 (sp)	100
CR50/125/250	50/125/250	Silica/silica/acrylate	0.20 (gd)	100
WF100/110P	100/110/135	Silica/silica/polyimide	0.22 (sp)	70
F-MCB-T	100/110/140	Silica/hard-polymer/tefzel	0.22 (sp)	100

sp = step index fibre; gd = graded index fibre.

Table 6.2 Commercial specifications of enamel-coated magnet wires

Wire type (gauge)	Measured resistivity (10^{-8} Ω·m)	Diameter of wire (mm)	Color of coating
22	1.3	0.67	Gold
26	1.9	0.42	Green
30	2.1	0.27	Red

shown in Fig. 6.11, are 3.5 mm for fibres F-MCB-T, WF 100/110P and FIP100110125; and 2.5 mm for fibre CR50/125/250. After all of the fibres have been optically and mechanically tested and the results analysed, it became clear that the materials of the fibre as well as the thickness of the cladding/coating are both important parameters in evaluating the critical bending radius of the fibre. These results were used to select the most compatible fibres for integration into textile structures (El-Sherif, 2002a).

Electric wires

Copper wires with a cylindrical shape are the most common wires used in electronic components. Therefore, samples of these wires were selected for integration with textiles; however, twisted and coaxial wires were also considered for future applications. Three types of enamel-coated wires (listed in Table 6.2) each having a different resistance, diameter and colour of coating were selected for integration into textile fabrics.

The resistances of these electronic wires have been measured at different bending curvatures (from 25 to 1.25 mm). There was no change in resistance for both the 30 gauge and the 26 gauge wires throughout the experiment on bending curvatures. However, the 22 gauge wire (the thickest) shows no change in resistance for all bending diameters larger than 5 mm. For a bending of less than 5 mm, the 22 gauge experiences a gradual increase in resistance, up to 50% at a bending radius of 1.25 mm (El-Sherif, 2002a).

6.4.2 Integration of optical fibres and electric wires into textiles

Several types of basic textile structures were used for the integration of optical fibres and electric wires. Such structures are warp knitting and weft knitting; plain, twill and satin weaving; and the non-woven fabrics produced by either heat bonding or adhesive bonding. The process of integrating optical fibres and wires into those textile structures would be fully automated in large-scale manufacturing. There are three requirements for a successful method of integration:

6.12 Optical image of a knitted fabric with embedded optical fibres.

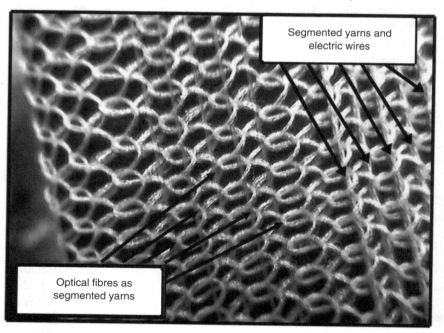

6.13 Closer look at optical fibre/wires integrated into knitted textile.

Integration of fibre optic sensors and sensing networks 121

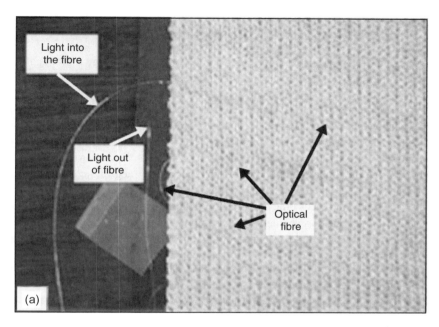

6.14 (a) Optical fibre integrated into a weft knitted textile in a serpentine shape and (b) a magnified part of (a).

1 The process of integration has to be applied to large-scale manufacturing.
2 The optical fibres and electrical wires should not show any signs of damage or signal weakness during the manufacturing process.
3 The application of the final product should not change the signal propagation more than the values indicated in the the predesigned conditions.

A study conducted recently has proven that the continuous integration of fibres/wires can be achieved through the ordinary operation of textile machines. Different types of selected optical fibres and electronic wires were successfully incorporated into knitted, woven and non-woven fabrics. Signal processing tests show that the developed methodologies have an almost negligible effect on the transmission of signals through the integrated fibres or wires.

Optical fibres and electronic wires can be integrated into knitted fabrics along the weft direction in short segments or continuously in a serpentine shape. Figures 6.12 and 6.13 show a sample of a weft knitted fabric with a number of optical fibres and electric wires integrated in the fabric as segmented yarns, and Fig. 6.14 shows a sample of the continuous feeding of the optical fibre in a serpentine shape. A specific time interval was set in order to create an acceptable serpentine shape of the bending curvature of the optical fibre (El-Sherif, 2002a). This method helped to avoid signal losses.

In order to insert an optical fibre and a wire into a woven textile, the fibre and wire must be woven into the structure just as the yarns are woven. In any weaving machine, the yarns in the warp direction are pulled out from a roller that contains many individual strands of yarns. These yarns are then formed into a woven structure by passing a strand of yarn in the transverse or weft direction via a shuttle, which interlaces the yarns. Significantly, the yarns in the weft direction are not continuous and their length cannot be very long since this length would match the width of the weaving machine. However, the yarns in the warp direction are much longer than those in the weft direction, since these yarns are continuously pulled out of a roller.

A very long piece of fibre/wire can be integrated into the fabric in the warp direction during the weaving process, as shown for the transverse yarn in Fig. 6.15 for a satin woven structure. In addition, the density of optical fibres and wires in the textile structure can be easily controlled. Therefore, in woven fabrics, optical fibres and wires are processed as warp yarns. The integrated fibre/wire took the exact same path as one of the yarns fed from the roller in the warp direction. Several samples were processed with fabric densities ranging from 25 to 200 PPI. The higher the density, the tighter was the structure of the fabric. A sample of a plain-woven fabric at 80 PPI is shown in Fig. 6.16. In this fabric, two optical fibres and two electric wires were integrated into the warp direction. Integrated fibres/wires interlace with a strand of one-weft yarn once with every strand of seven-weft yarn. The integration of fibres/wires did not influence the regular weaving procedure or the structure of the fabric because the integrated fibre/wire was treated as the

Integration of fibre optic sensors and sensing networks 123

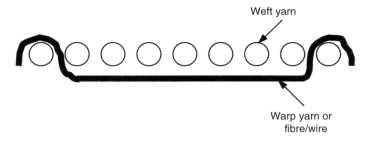

6.15 Schematic of cross-sectional 8-harness satin woven structure.

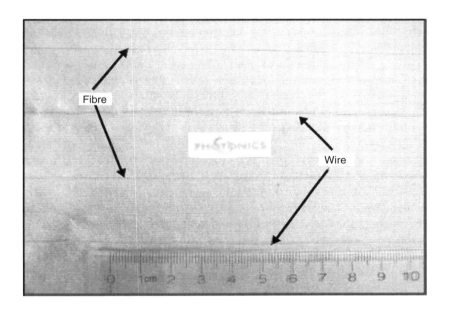

6.16 Different types of optical fibres and electronic wires integrated into a plain-woven structure at 80 PPI.

regular yarn in the warp direction. In the case of integrating fibres/wires into a nonwoven fabric, the fibres/wires were sandwiched between layers of fabric, as explained before.

The performance of the integrated fibres and wires was tested to evaluate the process of integration. The optical signals that had been transmitted through integrated fibres were measured and the resistance of the wires was tested. The results show that there was no change in optical signal loss or in wire resistance. Clearly, the integrated optical fibres were not excessively bent or stressed by the fabric yarns. In conclusion, the study thus proved that the methods of integration that had been developed are suitable for application with the commercially available textile machines. However, minor modifications are recommended for some of the machines (El-Sherif, 2003).

6.5 Applications of smart textiles

Because of the successful integration of optical fibres and fine electric wires into textile structures, a number of applications are under development. One of these applications is the smart parachute, which has the ability to predict the opening forces and to measure the deformation of the fabric of the parachute canopy during airdrops. Smart parachutes with embedded fibre optic sensors can be used for real-time characterisation of the dynamic structural behaviour of parachutes during airdrop (El-Sherif *et al.*, 2000a). *In situ* measurements of strain/stress in real time will permit better parachute designs in terms of structural parameters and selection of materials. Another application of smart textiles is the smart uniform, which has the ability to sense environmental conditions. The detection of biological or toxic substances is based on the concept of modifying optical fibres (passive conductors) to become chemically sensitive. This modification is achieved by replacing the passive cladding material in a small section of the optical fibre with a chemically sensitive agent (El-Sherif *et al.*, 2000b). When these sensors are incorporated into clothing (such as into uniforms for fire fighters, security guards and special mission personnel), they will provide instantaneous early warning of the presence of chemicals or toxins in the ambient environment (El-Sherif, 2001). A third application of smart textiles is the development of a smart shirt for health monitoring and diagnostics (Park and Jayaraman, 2001). This shirt can be used to monitor heart conditions, blood analysis and circulation, injury conditions, and other health issues. A large number of technical reports and papers have been published, recording the progress achieved thus far on these applications. In the next section, the application of smart textiles in parachutes is discussed.

6.5.1 Smart parachutes

Along with the idea on developing smart structures, an interesting research programme on the development of smart parachutes was carried out by a multidisciplinary team of researchers from the Fibre Optics and Photonics Center of Drexel University and Photonics Laboratories Inc., Philadelphia, Pennsylvania. Two types of fibre optic sensors were developed and applied for sensing static and dynamic loads in parachute canopy and suspension lines (El-Sherif *et al.*, 1999, 2001a). One of these types is the optical-fibre Bragg grating (FBG) sensor, which is used as a short-strain gauge for measuring axial strain fibres. The second type of fibre optic sensor that has been developed is based on measurements of modal power distribution (MPD) in multimode fibres. This sensor has the advantages of being inexpensive, highly sensitive, flexible and very small, which are the

Integration of fibre optic sensors and sensing networks 125

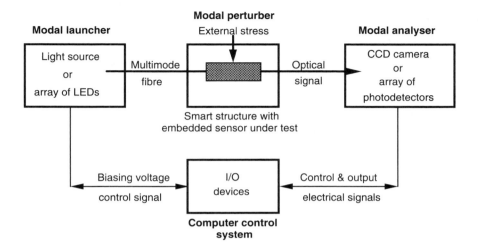

6.17 Function diagram of an application of a MPD sensor system.

requirements for sensors embedded in textile structures (El-Sherif, 1989; El-Sherif *et al.*, 1999). It can be used to measure axial as well as transverse strains. When integrated, these two types of sensors form a novel *in situ* monitoring system capable of measuring the dynamic structural behaviour of parachutes during inflation (El-Sherif *et al.*, 2001b). Since the application of FBGs as strain sensors is well known and has been reported elsewhere, the principle of operation of the developed MPD sensor will be presented next.

The principle of operation of the MPD technique that has been developed is based on modulating the modal power in multimode fibres. Within a multimode optical fibre, optical signals propagate according to the modal structure of the fibre and the boundary conditions. The MPD within a multimode fibre is a function of the geometry (size) and the optical properties (core and cladding indices) of the fibre and the light-launching conditions. Altering the boundary conditions of an optical fibre induces modal coupling, resulting in the modulation of the modal power distribution (MPD) (Radhakrishnan and El-Sherif, 1996). The Coupled Mode Theory can be employed for the analysis of the MPD modulation (Snyder, 1972; Snyder and Love, 1983).

Deforming the fibre by mechanical stress or other forms of perturbation results in the modulation of modal power, which can be exploited for sensing the applied signals. The measurements of the distribution and subsequent redistribution of the modal power can be accomplished by scanning the far-field pattern at the end of the fibre, using a charged coupled devices (CCD) camera or an array of photodetectors (Fig. 6.17). As an example, for an optical fibre with a core diameter of 20 μm, $n_{clad} = 1.45$ and $n_{core} = 1.46$, the field of the modal power for LP_{lm} modes

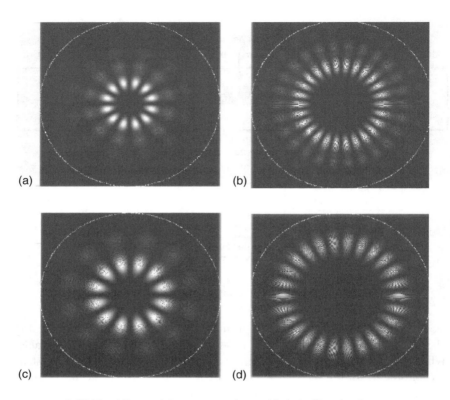

6.18 The LP_{lm} modal structure of a multimode fibre having; $n_{core} = 1.46$ and (a) $l = 6$, $n_{clad} = 1.45$, (b) $l = 13$, $n_{clad} = 1.45$, (c) $l = 6$, $n_{clad} = 1.455$, and (d) $l = 13$, $n_{clad} = 1.455$.

is shown in Fig. 6.18(a) and (b) for the modal orders $l = 6$ and $l = 13$, at an optical wavelength of $\lambda = 0.75$ μm. In the presence of external perturbation applied to the fibre resulting in a change of the cladding index to $n_{clad} = 1.455$, the modal power redistribution of $l = 6$ and $l = 13$ modes is shown in Fig. 6.18(c) and (d). This theoretical analysis is based on the use of a single-frequency (laser) light source. However, when a light-emitting diode (LED) is used, the modal power structure will have a continuous distribution of intensity. Through selective excitation, a limited number of propagating modes can be excited. This method can be applied by exciting the optical fibre with a beam of light off-axis.

For example, a step-index silica fibre with a diameter of 100 μm was excited at 10 degrees off-axis, using an LED. The two dimensional far-field pattern (MPD) and intensity profile were scanned and recorded by a CCD camera, as shown in Fig. 6.19(a). When the fibre was under stress, the recorded far-field pattern shows inter modal coupling and a redistribution of the modal power, Fig. 6.19(b). As the applied stress was increased, considerable rearrangement of the modal power was

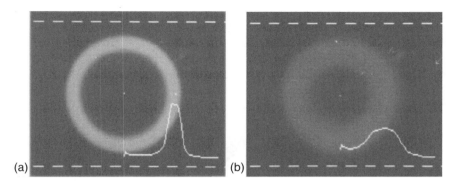

6.19 Far-field pattern and the intensity profile at the end of an optical fibre excited off-axis, (a) before and (b) after a stress was applied to the fibre.

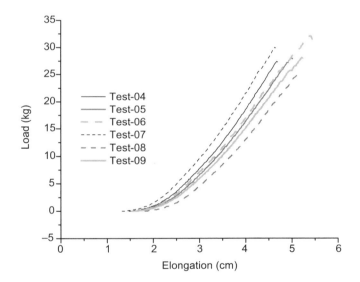

6.20 Tensile tester outputs for the six tested samples, for fabric align along 45 degrees, optical fibre stitched vertically.

recorded. These figures indicate that continuously varying the applied stress will result in continuous changes in the MPD.

For a simple approach and cost-effective sensor configuration, the CCD camera can be replaced by photodetectors located at key positions in the far field. Therefore, using an LED as the light source and regular photodiodes for detection will produce a sensitive, inexpensive and miniature sensor. These advantages make the MPD technique the most suitable one for use in parachute applications.

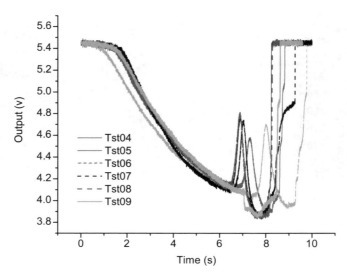

6.21 Optical sensors output for the six tested samples. MPD tensile test for fabric align along 45 degrees, optical fibre stitched vertically.

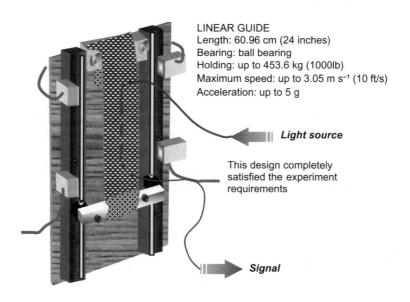

LINEAR GUIDE
Length: 60.96 cm (24 inches)
Bearing: ball bearing
Holding: up to 453.6 kg (1000lb)
Maximum speed: up to 3.05 m s^{-1} (10 ft/s)
Acceleration: up to 5 g

Light source

This design completely satisfied the experiment requirements

Signal

6.22 Experimental setup designed for free-falling drop tests.

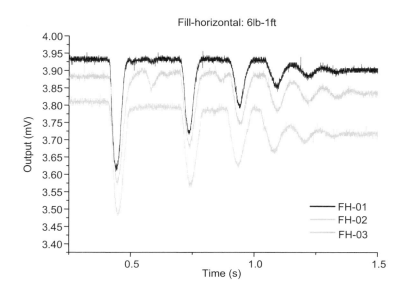

6.23 Drop test results of MPD sensors for three samples.

Experimental work was carried out to validate the MPD technique for measuring static and dynamic strains in textile structures. Quasi-static and dynamic tests were performed on processed samples. Strands of optical fibres were stitched into small pieces of a canopy nylon fabric and tested. The quasi-static test results are used to build the relationship between the optical signal and the mechanical behaviour of the fabric, and thus to predict the external perturbation on the smart fabric during dynamic applications (El-Sherif *et al.*, 2001a).

The quasi-static tests were carried out in a uniform, systematic way. The tensile tester pulled the fabric by a certain amount and the output data were then recorded for load, elongation and sensor output. This process was continued until the fabric began to tear. Figure 6.20 shows the tensile tester output (load versus elongation) for six tested samples, and the sensor output is shown in Fig. 6.21. It can be seen that the output of the sensor was in full agreement with the output of the tensile tester for the first 6 s, before the fabrics started to tear.

For dynamic tests, a test setup was designed to study the dynamic behaviour of the parachute fabrics and to develop the instrumentation needed for field applications. A drop test setup was designed, as shown in Fig. 6.22, for free-falling tests. Several tests were carried out using different weights and different dropping distances. In each case, the output of the photodetector was recorded with the use of an oscilloscope. Figure 6.23 shows the output of the photodetector, in terms of voltage versus time, for three samples tested at a weight of 2.72 kg (6 lb) and a drop height of 30.48 cm (1 ft). The sensor output for the three samples is identical. The

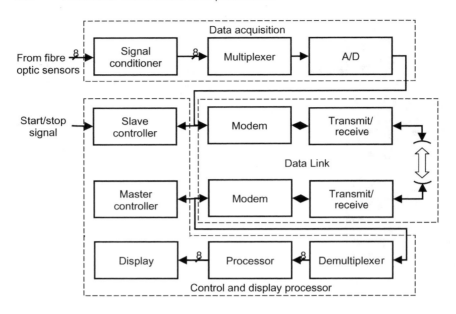

6.24 Schematic diagram of the eight channels RF remote data acquisition system.

first pulse records the output data for the drop test and the following pulses represent information on the oscillation. The tests were repeated for different weight and heights.

The major observation to be deduced from these data is that the height of the first pulse and the number of major pulses increase as the dropping distance increases. A second observation is that the delay time between the first two major pulses increases as the dropping distance increases. This increase in delay time reflects an increase in the time it takes for the load motion to bounce back for the second cycle of oscillation. As a result of these drop tests, a clear relationship exists between the dropping weight and dropping distance on the one hand and the pulse widths, time delay between the first and second pulses, and the number of pulses on the other. The number of experiments conducted was enough to draw the general conclusion on these relationships that the developed methodology can be applied to measure the dynamic strain during the time a parachute inflates. All of the recorded data can be directly translated to determine the amount of strain/stress applied to the fabric of the parachute during drop tests.

Based on the successful results achieved in the laboratory tests (quasi-static and dynamic) and on the mechanical analysis and modelling, a correlation function between the induced strain and the output of the sensors has been developed. Recently, a complete remote sensory system was developed and integrated into a small-scale parachute, and field tests were successfully carried out. The parachute used in the test was one-quarter of the scale of a regular personal circular

Integration of fibre optic sensors and sensing networks 131

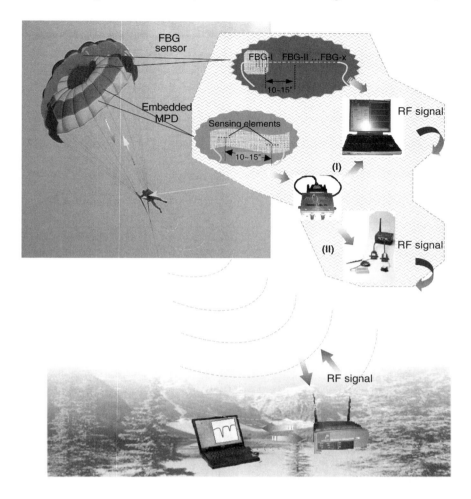

6.25 Schematic diagram of the developed smart parachute, showing the sensor's components, the RF transmitter receiver and the ground station.

parachute. A number of fibre optic sensors were integrated in the canopy of the parachute. Specifically, four sensors of each of the MPD and the FBG types were integrated in the radial and hoop directions of the canopy, two in each direction. For real-time monitoring of the sensor's outputs during the airdrop, an electronic device for data multiplexing was developed in connection with a RF wireless transmitter receiver, to transmit the data to a ground station, as shown in Fig. 6.24. A schematic diagram of the smart parachute with all the equipment and systems that had been developed is shown in Fig. 6.25. Field-drop tests have also recently been successfully carried out. The results have been presented in the Final

6.26 Scenes of the drop test, showing the parachute load where the sensor's data acquisition electronics and the RF transmitter are located.

Technical Report submitted by Photonics Laboratories, Inc. to the US Army under Contract # DAAD16-01-C-0003 (El-Sherif, 2002b). Various pictures captured during the drop test are shown in Fig. 6.26.

In conclusion, the remote sensory system that has been developed, embedded into the canopy of a parachute, has been proven to operate successfully during field tests. Although the remote sensor system was designed for *in situ* measurement of strain/stress in parachute canopies, it can be used to measure environmental conditions in spaces if the strain sensors are replaced by fibre optic chemical sensors. The methods developed to manufacture smart parachutes and smart uniforms can be applied to many other textile structures. Also, the methodology that has been developed presents excellent possibilities for the design of wireless fibre optic systems for health monitoring, not only in terms of textile structures but also in other structures and materials. Finally, the value of the MPD technique in textile-embedded sensor applications, where miniature and flexible sensor structures are of great importance, has been successfully demonstrated through these experiments on smart textiles.

6.6 Acknowledgements

Special thanks are accorded to the members of the research team who have worked with me for the past six years at the Fibre Optics and Manufacturing Engineering Center, Drexel University and Photonics Laboratories, Inc., Philadelphia, Pennsylvania, in developing various types of smart textiles. I am most grateful for the technical collaboration and continuous contribution of Jianming Yuan, Min Li, Dina El-Sherif, Mohamed Hidayet, Saif Khalil, Mohamed Abou-iiana, Fuzhang Zhao (for the development of the mechanical analysis), Rachid Gafsi, Kemal Fidanboylu, Lalit Bansal and Bulent Kose. Special appreciation also goes to Calvin Lee and James Fairneny of the US Army Natick Soldier Center, Natick, MA, for their continuous support and technical contribution.

6.7 References

El-Sherif M A (1989) 'On-fiber sensor and modulator,' *IEEE Transactions on Instrumentation and Measurements*, April, 595–598.

El-Sherif M A (1997), 'Fiber Optic Sensors for Soldiers' Smart Uniforms' (invited paper), *Third ARO Workshop on Smart Structures*, Virginia Polytechnic and State University, Blacksburg, Virginia, Aug. 27–29.

El-Sherif M A, Fidanboylu K, El-Sherif D, Gafsi R, Yuan J, Lee C and Fairneny J (1999) 'A novel fiber optic system for measuring the dynamic structural behavior of parachutes', *Fourth ARO (US Army Research Office) Workshop on Smart Structures*, State College, PA, August 16–18.

El-Sherif M A, Fidanboylu K, El-Sherif D, Gafsi R, Yuan J, Lee C and Fairneny J (2000a), 'A novel fiber optic system for measuring the dynamic structural behavior of parachutes', *J. Intell. Mater. Syst. Struct.*, **2**(5), 351–359.

El-Sherif M A, Yuan J and MacDiarmid A G (2000b), 'Fiber optic sensors and smart fabrics,' *J. Intell. Mater. Syst. Struct.*, **2**(5), 407–414.

El-Sherif M, Li M, El-Sherif D and Lee C (2001a), 'Fiber optic system for measuring the structural behavior of parachute airdrop: Quasi-static and dynamic testing, structure health monitoring', Fu-Kuo Chang (ed.), Stanford University, CRC Press, 733–741.

El-Sherif M, El-Sherif D and Lee C K (2001b), *Method and Apparatus for Evaluating Parachutes Under Load*, US Patent 6,299,104 B1.

El-Sherif M A (2001), 'The final technical report on "Sensors and Smart Fabrics",' *The MURI-ARO project on Functionally Tailored Textiles*, Contract #DAAH 04-96-1-0018.

El-Sherif M (2002a), *Manufacturing of Smart Textile Integrating Optical and Electrical Conductors, and Sensors, Phase I*, Technical Report, US Army Contract # DAAD16-01-C-0054.

El-Sherif M (2002b), *A novel fiber optic system for measuring the dynamic structuring behaviour of parachutes, Phase II*, Final technical report, US Army Contract, # DAAD16-01-C-0003.

El-Sherif M (2003), *Manufacturing of smart textile integrating optical and electrical conductors and sensors, Phase II*, Final technical report, US Army Contract, # DAAD16-10-C-0054.

Hearle J W S, Grosberg P and Backer S (1969), *Structural Mechanics of Fibers, Yarns, and Fabrics*, Vol. 1, Wiley-Interscience, New York.

Olofsson B (1964), 'A general model of a fabric as a geometric–mechanical structure', *J. Textile Inst.*, **55**(11), T541–T557.

Park S and Jayaraman S (2001), 'Adaptive and responsive textile structures (ARTS)', Tao X M (ed), *Smart Fibers, Fabrics and Clothing: Fundamentals and Applications*, Woodhead Publishing, Cambridge, England.

Peirce F T (1937), 'The geometry of cloth structure', *J. Textile Inst.*, **28**(3), T45–T96.

Radhakrishnan J and El-Sherif M A (1996), 'Analysis on spatial intensity modulation for fiber optic sensor applications', *J. Optical Fiber Technol.*, **2**(1), 114–126.

Snyder A W (1972), 'Coupled mode theory for optical fibers', *J. Optical Soc. Amer.*, **62**(11), 1267–1277.

Snyder A W and Love J D (1983), *Optical Waveguide Theory*, Chapman and Hall, London.

Zhao F, Abou-iiana M and El-Sherif M, 'Smart textiles integrating optical and electrical conductors – a fabric model for plain weaving', in press.

6.8 Bibliography

Buck J A (1995), *Fundamentals of Optical Fibers*, John Wiley & Sons, New York.

Collins G E and Buckley L J (1996), 'Conductive polymer-coated fabrics for chemical sensing', *Synth. Metals*, **78**, 93–101.

Croll R H, Klimas P C, Tate R E and Wolf D F (1981), 'Summary of parachute wind tunnel testing methods at Sandia National Laboratories', *7th AIAA Aerodynamics & Balloon Technical Conference*, San Diego, CA, Oct. 21–23, Paper No. 81-1931.

El-Sherif M A et al. (1990), 'Modal power distribution modulation for sensor applications', *Technical Digest of Annual Meeting*, Optical Society of America (OSA), Boston, Nov. 4–9.

El-Sherif M A and Yuan J (1999), 'Fiber optic sensors and smart fabrics', *Fourth ARO (US Army Research Office) Workshop on Smart Structures*, State College, PA, August 16–18.

El-Sherif M, Fidanboylu K, El-Sherif D, Gafsi R and Lee C (2000), 'A novel fiber optic system for measuring the dynamic forces in textiles, optical fiber sensors', *SPIE*, **4185**, 696–699.

El-Sherif M, Li M, Yuan J, El-Sherif D, Rahman A, Khalil S, Bansal L, Abou-iiana M, Lee C and Fairneny J (2002), 'Smart textiles with embedded opto-electronic networks and sensors – an overview', *International Interactive Textiles for the Warrior Conference*, Soldier Biological and Chemical Command, US Army Natick Soldier Center, Boston, Cambridge, MA, July, 9–11.

Garrard W L and Konicke T A (1981), 'Stress measurements in bias-constructed parachute canopies during inflation and at steady state', *J. Aircraft*, **18**(10), 881–886.

Garrard W, Konicke M L, Wu K S and Muramoto K K (1987), 'Measured and calculated stress in a ribbon parachute canopy', *J. Aircraft*, **24**(2), 65–72.

Jorgersen D S and Cokrell D J (1981), 'Aerodynamics and performance of cruciform parachute canopies', *AIAA 7th Aerodynamics & Balloon Technical Conference*, San Diego, CA, Oct. 21–23, Paper No. 81–1919.

Lee C K (1984), 'Experimental investigation of full-scale and model parachute opening', *AIAA 8th Aerodynamics & Balloon Technical Conference*, Hyannis, MA, pp. 215–223, April 2–4, Paper no. 84–0820.

MacDiarmid A G, Zhang W J, Feng J, Huang F and Hsieh B R (1998), 'Application of thin films of conjugated oligomers and polymers in electronic devices', *Polymer Preprints*, **339**(1), 82.

Rubber M F (1985), 'Polyurethane-diacetylene elastomers: a new class of optically active materials', *ACS Polym. Mater. Sci. Eng. Preprints*, **53**, 683–688.

Rubber M F (1996), 'Novel optical properties of polyurethane-diacetylene segmented copolymer', *ACS Polym. Mater. Sci. Eng.*, **54**, 665–669.

Saleh B E A and Teich M C (1991), *Fundamentals of Photonics*, John Wiley, New York.

Yuan J, Feng J, El-Sherif M and MacDiarmid A G (1998), 'Development of an on-fibre chemical vapor sensor', *OSA Annual Meeting*, Baltimore, MD, Oct. 4–9.

7
Wearable photonics based on integrative polymeric photonic fibres

XIAOMING TAO
The Hong Kong Polytechnic University, Hong Kong

7.1 Introduction

This chapter is intended to provide a review and overview of the development of wearable photonics with integrated polymeric photonic fibre structures. Photonic fibres can be defined as fibres that generate, transmit, modulate and detect photons. These fibres may form a base for a range of sensors as well as displays of flexible fabric that are capable of controlling colour, the intensity of luminescence, scattering intensity and self-amplification. Sensing by photonic fibres has been reviewed by the present author earlier (Tao, 2002); hence this chapter will concentrate on their display applications.

The generation and modulation of light within the visible range (380–780 nm) are the cause of the colour (wavelength) and intensity of materials. The major sources of colour can be grouped into the following categories: emissive radiation, absorption, reflection, dispersion, scattering and interference. In textiles, the most widely used mechanism is the absorption of dye stuff or pigments of certain wavelengths from white light. This is a subtractive modulation process. This method of coloration does not allow colour and intensity to be controlled after the process is completed. In other words, such a method produces a lifeless colour. Because of the wet chemical processes used, there has been pressure for the method to become more environmentally friendly.

7.2 Photonic band-gap materials

Photonic band-gap (PBGs) materials or photonic crystals (PhCs) are materials with a periodic dielectric profile, which can prevent light of certain frequencies or wavelengths from propagating in one, two or any number of polarisation directions within the materials. This range of frequencies is similar to an electronic band-gap; thus, it is often called a photonic band-gap. As shown in Fig. 7.1, the PBG materials can be one (1D), two (2D) or three-dimensional (3D). The Bragg

Wearable photonics based on integrative polymeric photonic fibres 137

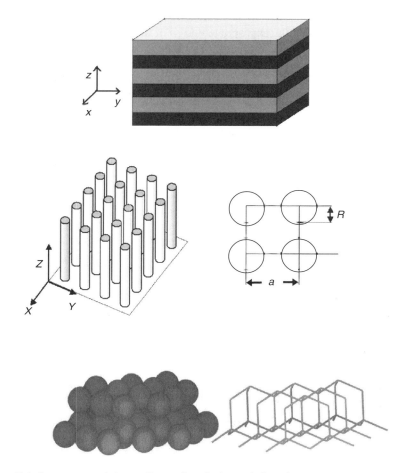

7.1 One, two and three-dimensional photonic band-gap structures.

grating structure is the best known one-dimensional PBG. Like an electronic band-gap, the PBG is caused by a lattice or a crystal structure. The lattice scale of PBG is in the order of the wavelength of light (0.1–2 mm), rather than in the order of atoms.

The PBGs may be passive or active. The interaction of passive or active PBGs with light can be described fully by using the Maxwell equations. Various numeric schemes have been developed and used to deal with these problems. We have used a finite difference time domain analysis to study the reflection and amplification of light in passive and active PGBs.

Light interacts with and within a PBG by multiple scattering and diffraction that is similar to Bragg reflection and diffraction gratings. Defects in a PGB will strongly affect the local electromagnetic field. Hence, these PGBs have been demonstrated to possess unique optical resonances and their properties can be tailor-made through the appropriate design of the band-gap and induced defects.

There are one, two and three-dimensional photonic band-gap structures. One-dimensional PBGs consist of alternating layers with different values of refractive index. Methods of fabrication include molecular beam epitaxy, chemical vapour deposition (CVD) technology, metallo-organic CVDs, metallo-organic vapour phase epitaxy and the holographic exposure of ultraviolet (UV) beams to photo-sensitive materials. Two-dimensional PBGs have periodicity along their two coordinate axes and homogeneity along the third. They can be fabricated through dry etching by reactive ions or wet electrochemical etching. The first method allows accurate control over the size and arrangement of holes (with nanometre precision), but has a limited maximum depth of etching. Wet electrochemical etching can obtain very deep holes and is suitable for fabricating high aspect ratio structures, but the size of the etched holes is somewhat less predictable.

Although two-dimensional PBGs provide some degree of three-dimensional control of the propagation of electromagnetic (EM) waves, truly three-dimensional PBGs are needed for full control via the effects of PBGs. Despite all the flexibility and precision of modern semiconductor processing techniques, experimental success in producing 3D PhCs is still somewhat limited. Among the methods of fabrication are: (1) colloid self-assembly; (2) the Yabonovite hole-drilling procedure using reactive ion etching; (3) etching and wafer fusion; (4) anodic aluminium oxide films; and (5) laser microfabrication.

7.3 Fibre-harvesting ambient light-reflective displays

Interference has been used by nature to produce many biological colours without the use of dyes and pigments. The wing scales of tropical Morpho butterflies form gratings that produce a brilliant blue colour by the constructive or destructive interference of light (Lippert and Gentil, 1959). The *Serica sericae beetle* and the indigo or gopher snake are other such examples. Reflection is the predominant mechanism in the colour of metals, which are a low-pass filter with a plasma frequency. Colour can also be produced by dispersion, or by the variation of the refractive index with a wavelength of incident light. All of these processes are passive, that is, they harvest and modulate ambient light.

Iridescent films and fibres have been produced commercially without dyes or pigments. Multilayered iridescent film is a thin film composed of a plurality of generally parallel layers of transparent polymeric material, in which the adjacent layers are of diverse materials whose refractive index differs by at least 0.03. The individual layers of film range from about 30 to 500 nm. Distinct colour reflections are obtained with as few as 10 layers, but for maximum colour intensity, it is desirable to have 70 or even more layers (Englehard Corporation, 2001). High colour intensity is associated with a reflection band that is relatively narrow and has high reflectance at its peak.

Optical fibres have been used for transmitting light and rapid technological advances have made the modulation and amplification of light by all fibre-based devices a reality. Various permanent grating structures (one-dimensional PBGs) have been developed by the modulation of the refractive index of the photosensitive core of germanium-doped silica fibres or polymeric optical fibres. The gratings have been used for optical filters/reflectors, fibre amplifiers, fibre lasers, sensors, and so forth. In the past, the author's group has investigated various methods of fabrication and characterisation for thin films and gratings of glass and polymeric fibres. Photosensitive polymeric single-mode optical fibres have been fabricated and characterised (Yu *et al.*, 2004). The team has also developed various new hybrid fibre-grating structures. In order to guide the design and fabrication of optical fibres and gratings, theoretical tools have been developed to deal with the optical responses (transmission/reflection spectra and polarisation states) of the optical fibres under various physical perturbations.

Holographic polymer dispersed liquid crystals (HPDLC) are a relatively new class of liquid crystal composites whose periodic structures are formed by holographic exposure to laser light. Phase separation occurs in the homogeneous mixture of liquid crystals and prepolymers during the laser irradiation induced polymerisation process, resulting in polymer-rich and liquid crystal-rich domains with predetermined periods. These structures are capable of reflecting or transmitting ambient light at controlled wavelengths. They are electrically controllable, self-adjusting in terms of reflected light intensity and have a high reflectivity of 70–100% (theoretical). Operating within the range of visibility, they produce tunable colours.

The interference of counter-propagating laser beams was first used in 1998 to create permanent switchable Bragg gratings consisting of a liquid crystal/polymer mixture sandwiched between two glass substrates. Electrically switchable Bragg grating (ESBG) generally refers to a multilayered HPDLC, whose transmission and reflection can be switched by applying electric voltage. The HPDLC is a variant of the conventional polymer dispersed liquid crystals that was developed for direct view, projection-display and switchable windows. The formation of the HPDLC Bragg gratings (a periodic structure with materials of an alternating refractive index) is based on the photopolymerisation of the photosensitive prepolymer under UV or visible light interferometric exposure. Figure 7.2 shows a SEM micrograph of an electrically switchable grating of HPDLC fabricated in the author's laboratory.

Compared with self-illuminating active displays, reflective or passive displays require relatively low operating voltage and are self-adjusting in reflective intensity. Reflective displays have many advantages over conventional liquid crystal displays. A higher reflectivity of between 70 and 100% can be achieved because no polarisers are used and thus the loss can be reduced (Crawford, 2003). The Bragg grating exhibits superior colour purity with narrow reflection peaks. This makes it possible to create red, green and blue reflectors without overlap of the

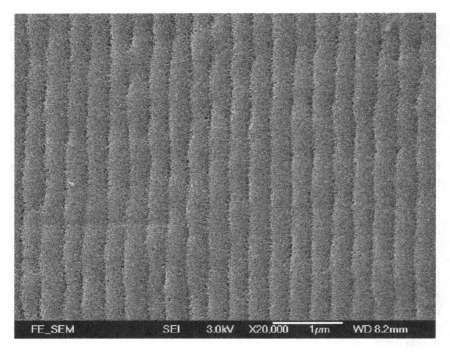

7.2 SEM micrograph of an electrically switchable grating of HPDLC fabricated in the author's laboratory.

reflection spectra, and to fabricate brighter full-colour displays with an extremely broad gamut of colour (Crawford, 2003; Bowley and Crawford, 2000). The optical properties of ESBG can be controlled by the action of an electric field. For example, the reflection peak can be tuned as a function of applied voltage. The reflective efficiency and peak wavelength of the HPDLC device can vary with those of other external fields such as shear-force. The ESBG or the HPDLC can be fabricated onto flexible substrates such as fabrics and films (Tomilin, 2003). The PDLC device can be easily manufactured over large areas and can achieve a fast rise time (~ 0.06 ms) and a high contrast ratio (~ 70:1) (for ferroelectric PDLC). The elimination of the backlight source may reduce power consumption and can make displays lighter and thinner.

7.4 Opto-amplification in active disordered media and photonic band-gap structures

One problem with current flexible displays based on side-emitting optic fibres is weak luminescence caused by the strong attenuation of light along the length of the fibres. Amplification along the length of the fibre is very desirable. One option is via opto-amplification.

The conjunction of the laser and the disorder medium has been of interest since shortly after the advent of the laser (Letokhov, 1967; Ambartsumyan *et al.*, 1970). In recent years various interesting interference effects have been recognised in light that is multiply scattered from disordered structures. There are several reference books on this topic: *Scattering and Localization of Classical Waves in Random Media* (Sheng, 1990), *Analogies in Optics and Micro Electronics* (Hareringgen and Lenstra, 1990) and *Introduction to Wave Scattering* (Sheng, 1995). For instance, it was found that the interference between counter propagating waves in disordered structures gives rise to enhanced backscattering. The phenomenon is known as coherent backscattering or weak localisation (Kuga and Ishimaru, 1984; van Albada and Lagendijk, 1985; Wolf and Maret, 1985). Later, more interference effects were recognised, such as the spatial correlations in the intensity transmitted through random media (Feng *et al.*, 1988). These experiments were performed on both active and passive random media. The action of amplification is explained by the Anderson location theory.

The multiple scattering of light is a common phenomenon in daily life, occurring, for example, in sugar, fog, white paint and clouds. The propagation of light in these media can in general be described by a normal process of diffusion. For the diffusion of light through a disordered material, the same Ohm's law holds as for the diffusion of electrons through any common resistor: the transmission or conductance decreases linearly with the length (thickness) of the system.

Anderson localisation brings classical diffusion to a complete halt. That is, on increasing the amount of scattering beyond a critical value, the material makes the transition into a localised state. Figure 7.3 shows the results of a simulation where localised rings appear with the lasing effect. This transition can best be observed in the transmission properties of the system. In the localised state, the transmission coefficient decreases exponentially instead of linearly with the thickness of a sample. (At the transition, the transmission coefficient is expected to have a power-law dependence on the inverse thickness, which is probably quadratic.) This will make backscattered light interfere and amplify in a passive medium.

Consider a light source in a disordered medium at position 'A'. A random light path that returns to the light source can be followed in two opposite directions. The two waves that propagate in opposite directions along this loop will acquire the same phase and, therefore, interfere constructively in 'A'. This leads to a higher probability that the wave will return to 'A' and, consequently, a lower probability that it will propagate away from 'A'. When the mean free path is reduced, the probability for such looped paths increases and at a sufficiently strong scattering, the system makes a phase transition from the normal conducting state into a localised state, owing to interference. In the localised regime, the system behaves as a non-absorbing insulator. Light that is incident on, for example, a slab or solution would be almost completely reflected and the remaining transmission would decrease exponentially with the thickness of the slab.

In this disordered medium, a dramatic narrowing of the spectrum, shortening of

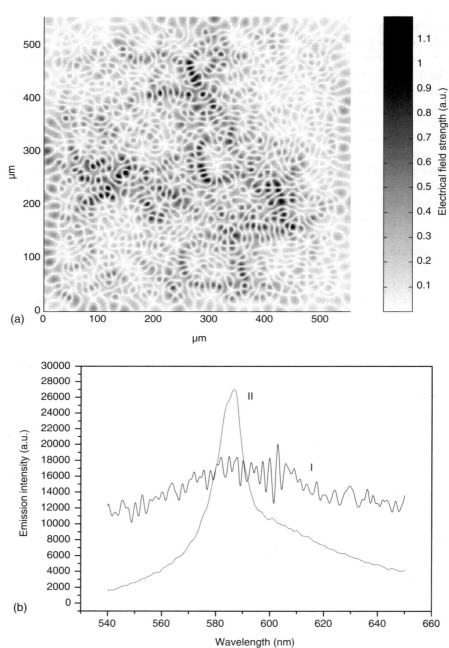

7.3 (a) Simulated E-field distribution where the localised rings appear with the lasing effect. (b) Emission spectrum of polymethyl methacrylate (PMMA) and TiO_2 nanocomposites (emission spectrum below (I) and above (II) the lasing threshold).

the emission time and abrupt buildup of peak intensity are observed above a threshold in pump energy. These results have raised the prospects of utilising the phenomenon for a variety of display, sensing and switching applications, particularly if the corresponding threshold can be significantly reduced. On the other hand, the non-saturable output and the narrow linewidth of the laser also open the door to a number of applications in medicine, identification and marking.

Over the past several years, there has been a great deal of renewed interest in multiple scattering systems with gain. In the case of gain, laser action has been demonstrated in these systems (Lawandy *et al.*, 1994; Diederik *et al.*, 1997) and has resulted in a number of technological applications. More recently, the effect of amplification on the coherent backscattering signal has been observed in a weakly amplifying sample of Ti_2O_3 doped Ti:sapphire powders and in a system with high-gain laser dyes and passive scatters. The amplification and temporal laser behaviour of light propagating in various types of disordered gain materials have been studied theoretically. But up to now, the exact mechanism of these observations is unknown and further experimental and theoretical investigations are required.

In a laser crystal powder medium, the amplification was ascribed to the diffusion process. Assuming a diffusion process, one can describe the time and position-dependent energy densities of pump, probe and amplified spontaneous emissions (ASE) by a diffusion equation with appropriate absorption and/or gain terms that depend on the local excitation of the system. The (also time-dependent) local excitation of the system is described by the set of rate equations of the laser material. The total set of coupled differential equations describing the system is formed by three diffusion equations for, respectively, pump light, probe light and amplified spontaneous emissions, and the rate equation for the concentration of laser particles in the metastable state.

In a nanoparticle doping dye medium, the physical picture of the system is as follows. The dye molecules absorb energy from the pumping beam and are excited to higher states. Then, through spontaneous emissions, some excited molecules randomly emit photons with frequencies different from that of the pumping beam. These photons travel in the medium, being scattered by the TiO_2 particles and amplified by the dye molecules through the stimulated-emission process. They finally leave the medium and reach the detector.

A laser model proposed by Lawandy and his colleagues sees the scatters providing feedback on the emitted light into the gain medium required to achieve laser action. This system has been modelled as a diffusive ring laser in a gain medium with distributed scatters. In this model the probability of a photon making a closed path in the gain medium by way of a random walk in the diffusive limit has been calculated and used to estimate the threshold gain. This model predicts the observed input–output characteristics and the emission linewidth as a function of pump energy.

The model is based on transient two-level laser equations and includes the detailed spectral properties of the dye gain system. The feedback mechanism in the laser model was quantified by a Monte Carlo simulation of this multiple scattering problem, which can be completely characterised by the scattering cross section and the Henyey–Greenstein phase function. The peak intensity theoretically calculated is in excellent agreement with the experimentally measured curves, whereas the linewidth data do not completely follow the experimental data at low energies, but collapse at the right energy for the densities of all scattering particles.

The disordered medium used can be classified into four categories. The first kind is a suspension of sub-micron polystyrene spheres. This is a kind of passive medium without gain. A high concentration is necessary to reduce the mean free path of light as much as possible, preferably down to its wavelength, which is necessary to induce the effects of location. The most effective scattering occurs if the size of the particle is of the order of the wavelength of light. The second is to use powder semi-conductor laser materials or to introduce scattering in laser materials, such as Ti:sapphire powders doped with Ti_2O_3 nanometre particles. These powders are fabricated by grinding laser crystals into powders or doping with scattering microparticles. The third is methanol solution doped with laser dyes and TiO_2 nanoparticles. The fourth is polymer sheets doped with laser dyes and TiO_2 nanoparticles. The latter three are active samples with a gain medium.

In general, the optical spectral properties of the above-mentioned disordered medium, including transmitting or backscattering light, are measured first to determine spectral narrowing above a certain threshold. Secondly, the peak input–output characteristics are given with a multichannel analyser and there is an obvious threshold, above which there is a linear increase. Thirdly, the influence of scattering particle density on the emission spectrum is also measured. Furthermore, the angular dependence and polarising properties of backscattering light in a passive medium are also involved in observations of good orientation emissions.

Photonic band-gap structures can suppress spontaneous emissions and control the lifetimes of chemical species in catalytic processes. If dye has infiltrated the PBGs, laser-like emissions may emerge in the mixture regime. It has been predicted that the lasing threshold can be reduced by introducing a defect into an otherwise periodic photonic band-gap structure. Since spontaneous emissions are suppressed in the band-gap, excitations will not then be drained by modes other than the lasing mode. In addition, the long dwell time of such localised defect modes reduces the gain required to reach the lasing threshold. Lasing has recently been observed in 2D photonic crystals, and promising 3D photonic crystals have been fabricated. The threshold for lasing may even be suppressed at defect modes in periodic structures that do not possess a full 3D photonic band-gap, such as in 1D periodic samples, including vertical cavity surface emitting lasers. Recently, lasing at the band edge has been demonstrated in dye-doped cholesteric liquid crystals (CLCs). In these chiral structures, a stop band exists for circularly polarised light that has the same sign of rotation as the CLC structure. Since the dwell time within the sample for emitted photons is enhanced near the edge of the band, the lasing threshold is also substantially reduced. There have been reports of other lasing dye-infiltrated PBGs. But almost all of these studies focus on how to prepare the PBG and characterise its structure and properties. In order to design and fabricate colloid self-assembled PBGs and their applications, it is necessary to investigate the relationship between the size and shape of the particle, lattice constants, material constants and the band-gap, the laser properties (intensity, wavelength) and dye concentration.

7.5 Electroluminescent fibres and fabrics

Emissive radiation, in which photons of specific energy and thus light with specific wavelengths are produced, may have many forms: black body radiation, fluorescence, photoluminescence, recombination of electrons and holes. All emissive radiations are additive. This kind of coloration has been widely used for lights with controlled colour and intensity of light, with television as a typical example. However, it has not been applied to flexible fibrous materials because of technical difficulties.

Recently, organic and polymeric electroluminescent displays have become an alternative to currently well-established display technologies such as cathode-ray tubes, liquid crystal displays (LCD) or inorganic materials. This is because such displays offer several advantages, such as self-light-emission, large viewing angles, low-driving voltages, high switching speeds, lightness of weight and solution processability. Conjugated polymers achieve semi-conducting properties owing to the existence of delocalised π-electrons along the polymer chain. Delocalised valence and conduction wavefunctions are formed containing π (bonding) and π^* (antibonding) orbitals, which facilitate the movement of mobile charges. Tables 7.1 and 7.2 show some semi-conductive materials being used in light-emitting diodes (LEDs). These organic/polymeric luminescent materials in a thin-film form are sandwiched between two electrodes. Injections of charge carriers, electrons and holes are then carried out in the emitting layer. They are consequently recombined and excited molecules are then formed. The molecules emit light once they decay. Therefore, the polymeric materials used in the device should possess high purity, be stable and easy to process, have good charge mobility with high quantum efficiency, and produce a band-gap corresponding to light emissions in a visible region.

Organic light-emitting devices (OLED) have been intensively investigated for use in flat panel displays since the first efficient OLED was discovered. OLEDs have been shown to have advantages over other display technologies, such as high brightness, low power consumption, low operating voltage, full viewing angle, high contrast ratios, fast response times and a wide temperature range. Compared to traditional OLEDs fabricated on glass substrates, flexible OLEDs (FOLED) are very lightweight, extremely rugged and more conformable, which makes them suitable for more user-friendly applications than can be managed by using flat panel displays, such as smart cards, wearable electronics, and so on.

Table 7.1 Molecular structures of typical semi-conductors

Commonly used acronym and chemical name	Chemical structure
Alq or Alq$_3$ *Tris*-(8-hydroxyquinoline)aluminium	
TPD *N,N'*-diphenyl-*N,N'*-bis- (3-methylphenyl)-(1,*N'*)-biphenyl- 4,4'-diamine	
PBD 2-(4-biphenyl)-5-(4-*t*-butylphenyl)- (1,1)-biphenyl-4,4'-diamine	
Bebq$_2$ *Bis*-(10-oxybenzo[h]quinolato)- beryllium	

Source: Friend, 1999.

7.5.1 Emitting species

To make highly flexible displays, the molecular bonds responsible for the mechanical properties of the thin films in the OLEDs must be tolerant of stress while the device is bent. The intermolecular bonding of organic molecules is a van der Waals force, the weak nature of which makes organic light-emitting materials suitable for fabricating FOLEDs. The structure of the OLED is given in Fig. 7.4. A layer of tris-(8-hydroxyquinoline) aluminium (Alq$_3$) was deposited onto an indium-tin oxide (ITO)-covered polyester sheet by vacuum-evaporation as the

Table 7.2 Conductive polymers used in polymer light-emitting diodes (PLEDs)

Commonly used acronym and chemical name	Chemical structure
PPV Poly(*p*-phenylene vinylene)	
MEH-PPV Poly[2-methoxy-5-(2'-ethylheoxy)-1,4-phenylene vinylene]	
MEH-CN-PPV	
PAT or P3AT Poly(3-alkylthiophene)	
PPyV Poly(*p*-pyridine)	
CN-PPV Poly[2,5-bis(hexylocy)-1,4-phenylene-(1-cyanovinylene)] ($R_1 = R_2 = C_6H_{13}$)	
PDPA Poly(diphenylacetylene)	

Table 7.2 Cont'd

Commonly used acronym and chemical name	Chemical structure
DHO-PPE Poly[1,4-(2,5-dihexoxy)-phenylene ethynylene] ($R_1 = R_2$ = hexyl)	
PPP Poly(p-phenylene)	
PDAF Poly(9,9-dialkylfluorene)	

Source: Nalwa, 2001.

emissive layer. Some other small molecules were also applied. The performance of these devices can be improved by addition of hole and electron-transporting materials, and can be compared with conventional OLEDs deposited on ITO-covered glass. The drop in light-emitting properties after repeated bending is miniscule. Thus far, Alq_3 has been intensively investigated in small-molecule flexible displays (Burrows *et al.*, 2001; Krasnov, 2002).

Although small-molecule organic light-emitting materials show potential in FOLEDs, their deposition on the substrate through vacuum-evaporation is rather complicated. Besides their high light-emitting efficiency, conjugated polymers have good flexibility and mechanical properties, which makes them suitable for FOLEDs. Moreover, the excellent film-forming property of the polymers makes it easy to fabricate the emitting films by spin coating. Derivatives of PPV were used to fabricate FOLEDs only two years after the electroluminescence of PPV was first achieved.

Gustafsson *et al.* (1992, 1993) described a flexible light-emitting device with poly(2-methoxy-5-(2'-ethyl-hexoxy)-1,4-phenylene vinylene) (MEH-PPV) as the emissive layer and poly(ethylene terephthalate) (PET) as the substrate. The device had a turn-on voltage of 2–3 V and an external quantum efficiency of about 1%. Its performance was almost the same as that of devices based on ITO-covered glass. Some other light-emitting polymers, such as polyfluorene and copolymers, were also used to fabricate FOLEDs for emissions of different wavelengths and high performance light-emissions (He and Kanicki, 2000; Paetzold *et al.*, 2003; Chang *et al.*, 2003). In these cases, the emitting polymers were all fabricated by spin-coating; however, the fabrication of film by spin coating has certain disadvantages,

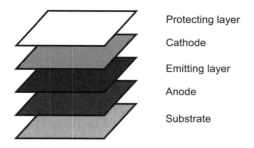

7.4 Structure of a typical LED.

such as the inefficient use of polymer solution and poor uniformity over large areas. Ouyang *et al.* (2002) developed a novel polymer thin film coating process, bar coating, to fabricate FOLEDs. Large area polymer thin films with a high uniformity were obtained, and high performance FOLEDs have also been produced with a performance comparable to devices produced by the spin-coating process. Although coating processes are convenient, they are still not ideal for the mass production of flexible displays. Bell Laboratory and some other groups are working on ink jet printing to produce polymer films or electronic circuits. The techniques show enormous promise – much lower cost production facilities, roll-to-roll rather than batch processing and opportunities to produce novel solutions such as flexible backplanes.

One of the challenges in FOLED displays is to achieve higher efficiency and minimise the consumption of power. Two types of excitons, singlet and triplet, are formed in a ratio of 1:3 in a working OLED. In a conventional fluorescent OLED, emission occurs only from the radiative decay of the singlet excitons; the radiative decay of the triplet excitons is forbidden, and the internal quantum efficiency is less than 25%. It was reported that phosphorescent emissions could be obtained from the radiative decay of the excited triplet states of phosphorescent organometallic materials, which are capable of producing an internal quantum efficiency of close to 100% (Adachi *et al.*, 2001). Weaver *et al.* (2002) fabricated flexible OLEDs with fac tris(2-phenylpyridine) iridium (Ir(ppY)3) as the phosphorescent dopant and 4, 4'-N, N'-dicarbazolebiphenyl (CBP) as the host; the doping concentration is 6%. This device emits green electroluminescence with a peak wavelength of 514 nm, an external quantum efficiency of 6.4% and a luminance efficiency of 22.7 cd A^{-1} at a current density of 2.5 mN cm^{-2}. Their lifetime can reach 3800 h from an initial luminance of 425 cd m^{-2}. Chwang *et al.* (2003) achieved phosphorescent FOLEDs with a lifetime of 2500 h for a 5 mm^2 FOLED pixel test. The encapsulated devices are flexed 1000 times around a cylinder 1 inch (2.54 cm) in diameter and exhibit minimal damage, showing potential for use in flexible displays.

7.5.2 Flexible substrates

The use of flexible substrates will significantly reduce the weight of flat panel displays and provide the ability to conform, bend or roll a display into any shape. Moreover, it will open up the possibility of fabricating displays by continuous roll processing, thus providing the basis for cost-effective mass production.

As a good barrier and transparent material, glass has been widely used in light-emitting devices. However it is brittle and can only sustain small strains, which limits its application in flexible OLEDs. Recently, Auch *et al.* (2002) reported their work on FOLEDs with ultrathin glass as the substrate and the encapsulating cover. The results show that when ultrathin glass is used and the OLED is kept at the neutral axis of the device, the flexibility can be considerable.

Metal foil has the excellent ability to prevent water and gas permeation. An efficient FOLED on a thin steel foil has been developed. Since light could not pass through the foil, light-emitting devices with such substrates have to be top-emitting. The device was constructed on the steel substrate with an organic stack of Alq_3 sandwiched by a highly reflective Ag anode and a semi-transparent cathode. It showed a peak efficiency of 4.4 cd A^{-1}, higher than the 3.7 cd A^{-1} of a conventional NPB/Alq_3 based OLED.

Polymers have excellent flexibility, are lightweight and low in cost, which makes them suitable for application in FOLEDs. So far, the most popular substrate used in FOLEDs is PET; however, it is permeable to water and oxygen, which will dramatically decrease the performance of the device. Polycarbonate and polypropylene adipate can also be severely damaged at high temperatures. Polyethersulfone is yellowish and absorbs moisture readily. Polyimide can bear high temperatures; however, it is also yellowish owing to intra- and/or intermolecular charge transfer complex formation.

To be effective in OLED applications, flexible films must demonstrate the following characteristics: low permeation rates from oxygen and moisture; a very smooth and uniform surface morphology; resistance to temperature and chemical use in conventional processing; optical clarity and transparency; and low cost.

7.5.3 Lifetime

The key challenge to developing FOLEDs is to achieve long-life operations. The failure mechanisms include fracturing of the ITO while bending and the permeation of oxygen and water. ITO is widely used as a transparent anode of OLED and other displays because it has high conductivity, high transparency and excellent workability. However, ITO has poor flexibility because of oxide ceramics. Chen *et al.* (2002) investigated the behaviour of these films under flexed conditions. The results show that a channelling crack is formed under tension, while under compression the film delaminates, buckles and cracks. Reducing the film or the thickness of the device will increase the allowable strain in the film. It is always possible to maximise the flexibility of the film by placing the most critical component near the neutral axis of the lamination.

It is now well known that the electronic properties of organic and polymeric materials used in FOLEDs or FPLEDs (flexible PLEDs) degrade rapidly in the presence of moisture and oxygen, and that encapsulation is necessary to achieve high performance flexible displays. A good barrier should combine good transparency, flexibility of the plastics and the excellent water/oxygen-preventing properties of an inorganic oxide. A device lifetime of 10,000 h requires a maximum leak rate of 5×10^{-6} g m^{-2}/day. The BARIXTM barrier was reported to reach this level. Chwang *et al.* fabricated high performance encapsulated FOLEDs with such barriers, and the lifetime of the device improved dramatically to as long as 2500 h.

7.6 Textile-based flexible displays

Wall tapestries and carpets are flexible displays that tell intricate stories of the universe, war, human life, etc. But they are lifeless and not interactive. The long-term targets of current endeavours are to make fabrics that can display pictures like a television screen, yet can be worn on the human body, folded and deformed. The fibre structures discussed in the previous sections are potential candidates for making such display devices, that is, light-harvesting fibres, light-emitting fibres and optical fibres.

So far, very few such display devices are known. Photo-adaptive fibres that can undergo photo-induced reversible optical and heat reflectivity changes have been developed. A US patent has described interweaving optical fibres to make side-emitting fabrics. Deflin *et al.* (2001) developed a way to perforate optical fibres with tiny holes that allow some of the light to escape sideways. There are prototypes for a flexible woven screen made of optical fibres capable of downloading and displaying static or animated graphics (such as logos, texts, patterns, scanned images, etc.) directly on to garments. The jacket can contain a very low resolution grid of eight by eight pixels, which displays crude yet readable symbols such as numbers. The author's team at The Hong Kong Polytechnic University has developed a number of flexible displays via various routes, as shown in Fig. 7.5.

Despite the technological advances and efforts of many teams around the world, truly flexible displays remain a dream at this moment.

7.7 Acknowledgements

The author wishes to acknowledge the funding extended by the Research Grants Council (Project No. PolyU5286/03E) and the Innovation and Technology Commission of the Hong Kong SAR Government (project No. ITS/071/02). The technical assistance of Dr. Sun XH and Dr. Huang HM in preparing the manuscript is also acknowledged.

(a)   (b)

7.5 (a) Flexible displays of the logo of The Hong Kong Polytechnic University. (b) Mannequin wearing fabric displays.

7.8 References

Adachi C, Baldo M A, Thompson M E and Forrest S R (2001), *J. Appl. Phys.*, **90**, 5048.
Akkermans E, Wolf P E and Maynard R (1986), *Phys. Rev. Lett*, **56**, 1471.
Ambartsumyan R, Basov N, Kryukov P and Letokhov V (1970), in *Progress in Quantum Electronics*, Sanders J and Stevens K (eds), Pergamon, Oxford, pp. 107–185.
Anderson P W (1958), *Phys. Rev. B*, **109**, 1492.
Auch M D J, Kian Soo O, Ewald G and Soo-Jin C (2002), *Thin Solid Films*, **417**, 47.
Balachandran R M and Lawandy N M (1994), In *Proceedings International Quantum Electronics Conference*, Vol. 9 of 1994 OSA Technical Digest Series, Optical Society of America, Washington, DC, p. 239.
Balachandran R M, Lawandy N M and Pacheco D P (1994), Presented at the 1994 Optical Society of America Annual Meeting, Dallas, TX, 2–7 October 1994.
Bowley C C and Crawford G P (2000), 'Improved reflective displays based on polymer-dispersed liquid crystals', *J. Opt. Technol.*, **67**(8), 717–722.
Burroughes J H, Bradley D D C, Brown A R, Marks R N, Marckay K, Friend R H, Burns P L and Holmes A B (1990), *Nature*, **347**, 539.

Burrows P E, Graff G L, Gross M E, Martin P M, Shi M K, Hall M, Mast E, Bonham C, Bennett W and Sullivan M B (2001), *Displays*, **22**, 65.
Chang S M, Su P K, Lin G J and Wang T J (2003), *Synth. Met.*, **137**, 1025–1026.
Chen Z, Cotterell B and Wang W (2002), *Eng. Fracture Mechanics*, **69**, 597.
Chwang A B, Rothman M A, Mao S Y, Hewitt R H, Weaver M S, Silvernail J A, Rajan K, Hack M, Brown J J, Chu X, Moro L, Krajewsk Ti and Rutherford N (2003), *Appl. Phys. Lett.*, **83**, 413.
Crawford G P (2003), 'Electrically switchable Bragg grating', *Optics & Photonics News*, 54–59.
de Boer J F, van Albada M P and Lagendijk A (1992), *Phys. Rev. B*, **45**, 658.
Deflin E, Weill A and Koncar V (2001), 'Communicating clothes: optical fibre fabric for a new flexible display', *Proceedings of ATC-6*, Hong Kong.
Diederik S W, Wiersma D S, Bartolini P, Lagendijk A D and Righini R (1997), *Nature*, **390**, 671–673.
Engelhard Corporation (2001), United States Patent no. 6291056.
Feng S, Kane C, Lee P A and Stone A D (1988), *Phys. Rev. Lett.*, **61**, 834.
Friend R H (1999), *Nature*, **397**, 121–128.
Gu G, Burrows P E, Venkatesh S and Forrest S R (1997), *Opt. Lett.*, **22**, 172.
Gustafsson G, Cao Y, Treacy G M, Klavetter F, Colaneri N and Heeger A J (1992), *Nature*, **357**, 477.
Gustafsson G, Treacy G M, Cao Y, Klavetter F, Colaneri N and Heeger A J (1993), *Synth. Met.*, **57**, 4123.
Hareringgen W V and Lenstra D (1990), *Analogies in Optics and Micro Electronics*, Kluwer Academic, Boston, MA.
He Y and Kanicki J (2000), *Appl. Phys. Lett.*, **76**(6), 661.
John S (1995), In *Confined Electrons and Photons*, Vol. 340 of *NATO Advanced Study Institute, Series B: Physics*, Burstein E and Weisbuch C (eds), Plenum, New York, 523.
Krasnov A N (2002), *Appl. Phys. Lett.*, **80**(20), 3853.
Kuga Y and Ishimaru A (1984), *J. Opt. Soc. Am. A*, **8**, 831.
Kwan C C, Tao X M and Sung X H, in preparation.
Lawandy N M, Balachandran R M, Gomes A S L and Sauvain E (1994), *Nature*, **368**, 436.
Letokhov V (1967), *Zh. Eksp. Teor. Fiz.*, **53**, 1442 [*Sov. Phys. JETP*, **26**, 835 (1968)].
Lippert W and Gentil K Z (1959), *Morph Okol. Tiere*, **48**, 115.
Nalwa, H S (2001), 'Electroluminescence in conjugated polymers', in *Handbook of Advanced Electronic and Photonic Materials and Devices – Light-emitting Diodes, Lithium Batteries and Polymer Devices*, Nalwa H S (ed.), Academic Press.
Ouyang J, Guo T, Yang Y, Higuchi H, Yoshioka M and Nagatsuka T (2002), *Adv. Mater.*, **14**, 915.
Paetzold R, Heuser K, Henseler D, Roeger S, Wittmann G and Winnacker A (2003), *Appl. Phys. Lett.*, **82**(19), 3342.
Sheng P (1990), *Scattering and Localization of Classical Waves in Random Media*, World Scientific, Singapore.
Sheng P (1995), *Introduction to Wave Scattering*, Academic Press, San Diego, CA.
Tang C W and VanSlyke S A (1987), *Appl. Phys. Lett.*, **51**, 913.
Tao X M (2002), *Smart Fibres, Fabrics and Clothing*, Woodhead Publishing, Cambridge, UK.
Tao X M and Sirikasemlert A (1999), *Text. Res. J.*, **69**(1), 43–51.

Tomilin M G (2003), 'Advanced display technologies', *J. Opt. Technol.*, **70**(7), 454–464.
van Albada M P and Lagendijk A (1985), *Phys. Rev. Lett.*, **55**, 2692.
Wang H M, Tao X M, Newton E and Che C M (2002), *Polym. J.*, **34**(8), 575–583.
Wang H M, Tao X M and Newton E (2004), *Polym. Internat.*, **53**(1), 20–26.
Weaver M S, Michalski L A, Rajan K, Rothman M A, Silvernail J A, Brown J J, Burrows P E, Graff G L, Gross M E, Martin P M, Hall M, Mast E, Bonham C, Bennett W and Zumhoff M (2002), *Appl. Phys. Lett.*, **81**, 2929.
Witt A N (1977), *Astrophys. J. Suppl. Ser.*, **35**, 1.
Wolf P E and Maret G (1985), *Phys. Rev. Lett.*, **55**, 2696.
Yu J M, Tao X M and Tam H Y (2004), *Optics Lett.*, **29**(2), 156–158.

8
Communication apparel and optical fibre fabric display

VLADAN KONCAR,
ENSAIT-GEMTEX Laboratory, France
EMMANUEL DEFLIN and ANDRÉ WEILL
France Telecom Recherche et Développement, France

8.1   Introduction

In the first part of this chapter we introduce basic definitions of communication apparel, describing the process of conception and its main components. Building blocks that have to be used in order to realise these generations of apparel are mentioned and analysed from the point of view of textiles. A classification of innovative, communicative and intelligent functions attributed to communication apparel is also developed.

When analysed individually, the terms 'apparel' and 'communication' indicate well-defined meanings that are closely related to our style of living and our environment. Therefore, our 'apparel' defines our preferences, style or social position, and our 'communication' defines the way we communicate with people, including our family, friends and professional colleagues.

What does the expression 'communication apparel' convey? What kind of communication function is supposed to be integrated into apparel and what is the purpose of this integration? We could say that fashion and clothing styles may explain the basically passive communication expressed by the apparel, for instance colour and patterns.[1] However, this passive communication function of apparel is not the focus of this study. We shall focus instead on the active communication that has to be implemented into apparel in order to enhance the traditional and well-known functions of clothing.

Thus, new forms of communication integrated in apparel, based on the latest electronic devices and data transmission processes, are examined. These should clarify the possibilities for the use of new technology and indicate probable future development related to communication apparel. The area to be investigated is vast and concerns various applications in the fields of health care (e.g. vital functions monitoring, diagnostics, etc.), military applications (combat apparel, injury detection, etc.), leisure (games, multimedia, etc.), business and many others. Nevertheless, all of these applications have a common point – the integration of electronic systems into textile supports.

After covering this step, we will analyse various emerging technologies which allow the integration of electronic devices into garments and textile accessories. Finally, it is important to note the distinction between wearable communication and 'wearable computers', which are not incorporated into the clothing itself, but transported as objects. Wearable communication also differs from 'intelligent clothing', which reacts to exterior or physiological stimuli to regulate and control the user's well-being, like the vitamin C distributing T-shirt, for example. In the second part of this chapter, a new approach to the design of flexible textile displays, simplifying the concept and overcoming the drawbacks of liquid crystal display (LCD) or cathode ray tube (CRT) video screens (rigidity, volume and weight), will be proposed. In addition, a display based on fabric made from optical fibres and classic yarns is described. The fibres are first specifically processed so that light can scatter throughout the outer surface of the fibres, resulting in a transversal restitution of the light. The diameter of the fibre section can be reduced to 0.25 mm, enabling extremely thin and light fabrics to be developed.

8.2 Communication apparel

Initially, and from a purely technical point of view, the concept of communication apparel may be perceived as the result of a convergence of two industries: textiles and electronics. The miniaturisation of electronics makes it possible for people to carry with them all kinds of devices, qualified as 'portables', with functions ranging from leisure (Walkman, MP3, portable television), communications and information management (mobile phones, personal digital assistants) to health (pacemakers, physiological sensors of parameters).

In another area, the textile industry has made considerable strides in the field of high value-added textiles, mainly in the sectors of high-performance textile and fibres. The use of new materials and the development of new structures and integration processes make it possible to develop supports able to convey information while being mostly based on the properties of electric conduction. These new achievements in the textile industry enable electronic devices to be directly integrated into the structure of textile, therefore modifying the functionality of the apparel. Besides the main functions of apparel, which are protection and passive communication, the clothes become the second skin[1] or an interface with specific functions between the individual and his or her environment.

8.2.1 From basic clothing to new communication apparel

It seems appropriate to examine first the traditional functions of clothing in order to be able to detect the latent needs that could be met by adding intelligent and

Communication apparel and optical fibre fabric display 157

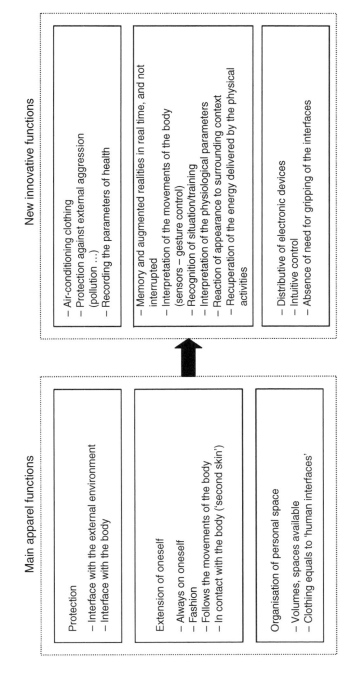

8.1 Intrinsic functions of clothing and new concepts.

communicative functions. Figure 8.1 summarises those traditional functions in three main areas: protection, extension of oneself and organisation of personal space.

More and less futuristic projections are given in the second part of this figure on what new functions the clothing of tomorrow could have. All of these concepts can be linked to the intrinsic functionality of clothing. The majority of these new functions are technically feasible today, but require the contribution of various technological specialities (textiles, electronics, telecommunications) according to the clothing's so-called 'intelligence level'. A classification of these new types of clothing is proposed in the next section.

Innovative high-performance textile fibres, yarns and fabrics, combined with miniaturised electronic devices, enable several intelligent functions to be incorporated into apparel. Figure 8.2 shows the stages of development that lead to the elaboration of communication apparel, including the inclusion of telecommunications functions. This figure also classifies apparel in terms of intelligence and communication. All of the technologies used in the process of elaboration (high-performance textiles, electronics, communications and telecommunications) are related to blocs describing the properties that can be useful in the conception of communication apparel. All of these technologies add new functions to the communication apparel, leading to changes in the way we define this apparel. The properties of intelligent and communication apparel, and their potential targets and applications, are detailed in the next sections.

8.2.2 Intelligent apparel

The term 'intelligent apparel' describes a class of apparel that has active functions in addition to the traditional properties of clothing. These novel functions or properties are obtained by utilising special textiles or electronic devices, or a combination of the two. Thus, a sweater that changes colour under the effect of heat could be regarded as intelligent clothing, as well as a bracelet that records the heart rate of an athlete while he or she is exercising. Intelligent clothing can therefore be classified into three categories:[1]

- clothing assistants that store information in a memory and carry out complex calculations
- clothing monitors that record the behaviour or the health of the person
- regulative clothing, which adjusts certain parameters, such as temperature or ventilation.

Finally, all intelligent clothing can function in manual or automatic mode. In the case of manual functioning, the person who wears the clothing can act on these additional, intelligent functions, while in the automatic mode the clothing can react autonomously to external environmental parameters (temperature, humidity, light).

Communicative clothing can be perceived either as an extension or as the next

Communication apparel and optical fibre fabric display

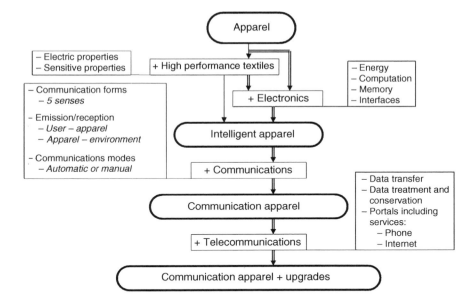

8.2 Apparel evolution.

generation of intelligent clothing. Although all clothing communicates intrinsically by virtue of its appearance, the type of communication referred to here is that of information coded and transmitted by means of electronic components in the clothing. In addition to the first examples of the integration of portable telephones and miniature PCs, many applications are being studied and have yet to be imagined. Communication can indeed be achieved between the clothing and the person who wears it, or between the clothing and the external environment and other people. In both cases, 'communicative' clothing refers to any clothing or textile accessory that receives or emits information to or from the structure that composes it.

8.2.3 Potential targets and applications

Everyone wears clothing, and most people are concerned with the appearance of communication apparel. However, needs will be different within any given group of people. Let us simply note that the broad, principal topics are:

- professional[2,3] (the need for 'free hands' functions, safety, data exchanges)
- health care[4] (monitoring, training, remote diagnosis)
- everyday life[5] (telephony, wellness)
- sports[6,7] (training, performance measurement)
- leisure (aesthetic personalisation, network games).

8.2.4 Technical elements enabling production

Previous sections have described communication apparel as an extension of the functionality of intelligent clothing. A study of the various technologies involved in the process of producing intelligent clothing can help to anticipate the new uses and new communication services that could be added to clothing. It is therefore advisable to have a vision of the various techniques likely to confer an unspecified form of intelligence on clothing.

In Fig. 8.3, the various stages concerning the design of new communication apparel are shown with building blocks (peripherals, processing data, connectors and energy) that have to be utilised in order to realise new specific functions. These building blocks are obtained using electronic components or textile electronics combined with technological and integration processes. Finally, the building blocks for several important properties are given. In the following sections the building blocks for integration are developed.

8.2.5 Various building blocks for integration

The second column of the table in Fig. 8.3 shows the classification of various electronic parts that can be included in communicative clothing, according to four principal recurring topics: peripherals, processing data, connectors and energy. A short description of these components is provided below, in order to understand better the objectives of research currently being undertaken on electronics 'related to oneself'.

8.2.6 Peripherals

The main peripherals supposed to be used in communication apparel are mentioned and quickly analysed in next few paragraphs.

Control interfaces – near the 'human interfaces'

The use of clothing to support control interfaces is interesting because the control interfaces can be close to the parts of body that are concerned,[8,9] for example, earphones in a collar or a bonnet, a microphone in a collar or a keyboard applied to the sleeve of a jacket. Another interesting example is, of course, voice recognition.[10–13]

The ergonomic adaptation to clothing of all of these control interfaces is also very important. In contrast to certain miniaturised communicative devices, clothing has a greater surface area, which enables it to offer more functionality. For example, the small keyboard of a mobile phone that fits in the palm of one's hand becomes much more readable when transposed to the surface of a piece of clothing that is three times larger. On the other hand, the lightness and flexibility that also

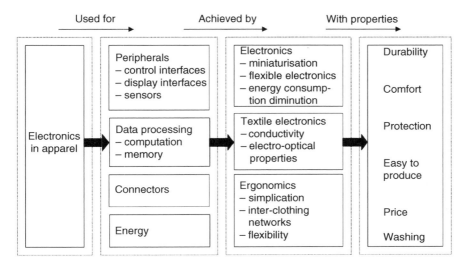

8.3 Technology and constraints of electronics integration into apparel.

characterise clothing implies a need to redefine the forms and materials employed for these new interfaces. New properties guaranteeing resistance to wear and to washing must also be taken into account.

Sensors

Since clothing accompanies every body movement and is sometimes in direct physical contact with the person, it has become an ideal physical support for translating and interpreting human activity by means of sensors. Clothing could be used to detect different actions, in particular the recognition of gestures, in order to facilitate certain commands that are intuitive, as with the automatic release of a phone call when one moves the collar of clothing to the ear.[14,15] Moreover, when these sensors are associated with computing and with the control unit, they may allow the recognition of situation and context for a better interpretation of reality.

Sensors in communicative clothing could also be used as psychological sensors for various parameters. This term refers to the sensors used to record health or person parameters in a broad sense. The applications rising from the use of these sensors are numerous. We can, for example, use sensors to provide a physical performance analysis of an athlete, or to conduct a patient medical follow-up in real time.

Interfaces of information restitution

In many applications, it is necessary to display or reproduce the information produced by communicating systems integrated into clothing. Therefore, traditional interfaces such as displays, screens or loudspeakers have to satisfy the same ergonomics and mechanical resistance criteria as those quoted in the case of control interfaces. Concerning colour liquid crystals screens, for example, the aspects of rigidity, weight and consumption, which characterise them at the present time, have to be adapted. Solutions containing microscreens in glasses or using technologies, including flexible supports, have begun to appear.

In addition, the proximity of clothing and textile accessories to the natural human senses opens new possibilities for the transmission of information. Visual and auditory ways of collecting information (such as screens and loudspeakers), which are today largely developed because they do not require direct contact with the user, could soon be joined by tactile and olfactory methods. The T-shirt with a collar that translates environments by diffusing a combination of perfumes is about to leave the realm of science fiction.

8.2.7 Data processing

The material supports of memory, computation and data processing (RAM, hard disks and processors) will certainly not evolve much in the short term unless they do so in the direction of miniaturisation. Even if developments are achieved on flexible substrates, they remain fragile and require partly rigid protection in order to be integrated into communication apparel. However, their integration has become entirely possible, as seen in the incorporation of a micro PC into the loop of a belt. It is also possible to imagine that only a small quantity of information could be processed locally in communicative clothing, while more complex functions and more significant memory capacities are handled by higher powered remote servers. This difference between local and mass treatment involves the development of specific algorithms, as is the case for intelligent vehicles.

8.2.8 Connector industry

Connection problems are another major issue in state-of-the-art communicative clothing. The principal question is how to transport information and energy between the various components of the electronic system with optimal efficiency. The concepts of weight distribution and ergonomics must be taken into account in distributing the various components on various zones of the body.

Diverse techniques of wireless transmission exist; for example, infrared or radio operator waves using various standards (IEEE 802.11, Bluetooth). If these modes of transmission are to free communicative clothing from the need for physical connections, several additional constraints must be taken into account. For example, the energy consumption necessary for their operation may be important. Moreover, when it is a question of simple information transport (such as an open or closed contact or something similar) or of energy transport, wired connections become indispensable. The wireless connections mainly have to be used to connect the user to the external environment. In addition, it seems interesting to have only one energy source distributed to the disparate electronic interfaces, thus allowing better energy management. On the other hand, each electronic interface could have its own computation and storage capacity, which would allow resources to be allocated and weight to be distributed.

It is important to examine the problem of control and the centralisation of information restitution. In fact, to be able to manage all of the functions of a complex communicating device, it is necessary to centralise outgoing controls and incoming information on a single interface. This means that accessing emails or finding a direction on a cartographic site, for example, must be done on a single screen.

8.2.9 Energy

Autonomy in energy is still a main handicap of the majority of mobile electronic devices. Many users of wireless devices have no doubt dreamt of never having to reload their mobile phones. Even if electronic circuits require increasingly less energy, new possibilities appear and create an additional need for energy (a larger screen size implies a need for greater power consumption).

Even in the case of communication apparel, autonomy versus weight and volume is once again a compromise that must be made. Battery technologies evolve (e.g. lithium–polymer) but, unfortunately, the batteries are still often the heaviest part of portable devices. The advantage of communication apparel is that the weight distribution in clothing will make it possible to be partly freed from this constraint.

Another interesting alternative seems to be the use of renewable energy sources. Solar energy and wind are relatively poorly adapted to clothing because they require large surface areas to be truly effective. On the other hand, many studies have been carried out on techniques that will make it possible to recover the energy released by the physical activity of the human body during the day. And, once more, clothing is an ideal support for these new recharging systems.

8.3 Optical fibre fabric display

Several different projects dealing with flexible displays and screen development have been carried out over the past decades. The final objective is to obtain sufficiently bright and flexible displays in order to facilitate their integration into communicative clothing. Different approaches have been developed involving new textile materials or using the optical fibres in the textile structures. These approaches are discussed in the next sections.

8.3.1 Textile-based flexible displays

There are several approaches to textile-based displays. The research project developed at Auburn University[16] deals with photo-adaptive fibres for textile materials. Moreover, the aim of this project is to develop photo-adaptive fibres that can undergo photo-induced reversible optical and heat reflective changes. Early on, thin and optically transparent polymer films were prepared to study the kinetics of particle evolution occurring in photosensitive fibres. The films were optimised for speed in metal particle formation and were prepared exclusively at high light intensities. These films will be used to study the chemistry of interfacial regions, which seem to have similar properties to the fibres. This approach will then be generalised to produce photo-adaptive fibres in order to make flexible displays using this type of fibre.

Another very interesting research project, in the field of 'chameleon fibres', has been developed at Clemson University's School of Textiles.[17] The aim of this project is to create modifiable colour fibres and fibre composite structures. This is supposed to be accomplished by incorporating molecular or oligomeric chromophoric devices capable of changing colour over the visible portion of the electromagnetic spectrum into (or onto) fibres. This is done by the application of a static or dynamic electrical field. Deliverables envisioned for this type of material include wall and floor coverings that change colour and also 'smart' and communicative clothing with flexible displays. Research on this subject has been conducted in a complementary manner in the laboratories of Furman University, Clemson University and the Georgia Institute of Technology.[18] Colour change is due to the absence of specific wavelengths of light, which, owing to structural changes, will vary with the application of an electromagnetic field.

Electrically conductive fibres can be used to provide a source for generating the electrical field necessary for colour change. Films also have the potential to be applied directly as coatings or polymerised directly on fibre or textile substrates by *in situ* processes.[19] The electrical field strength necessary to bring about dynamic colour change will depend on the choice of oligomer or molecular species, either attached to the fibre or to the surface of the film or embedded within the matrix of the material. A colour change from green to light blue has already been demonstrated for a film containing an oligomeric species in a small applied electric field.

The Visson company has also recently developed display prototypes based on a 0.2 mm thin textile fabric.[20] The display is an assembly of wire conductors woven in an X–Y structure, in order to create a rows-and-columns electrodes network. Each one of these conductive fibres is covered with a very fine layer of electroluminescent material. By addressing an electric voltage to one column and one row simultaneously, the electric field created at the intersection of the corresponding fibres causes electroluminescent material to be emitted at this point.

Some interesting studies also deal with nanocomposite fibres that could be used to develop the flexible displays. The project is the development of biphasic fibres with properties that leapfrog those of the matrix polymers. For example, improved high-temperature mechanical performance, useful optical properties and electrical or barrier properties of these fibres will have a major impact on titre reinforcement, electro-optical devices and other applications.[21]

8.3.2 Optical fibres in textiles

Optical fibres are currently being used in textile structures for several different applications. They are first often used as sensors exploiting the Bragg effect. At The Hong Kong Polytechnic University, Tao has developed several very important applications using optical fibres to measure strain and temperature in composite structures.[22-25] These fibre optic sensors have also been used in smart textile composites.[26] Actually, fibre optic Bragg grating sensors are attracting considerable interest for a number of sensing applications[27, 28] because of their intrinsic and wavelength-encoded operation. There is great interest in the multiplexed sensing of smart structures and materials, particularly for the real-time evaluation of physical measurements (e.g. temperature, strain) at critical monitoring points. In order to interrogate and demultiplex a number of in-fibre Bragg grating sensors, whether or not they are in a common fibre path, it is essential that the instantaneous central wavelength of each sensor can be identified.

The research team of Jayaraman[29] at Georgia Tech developed a smart shirt called the Georgia Tech Wearable Motherboard™ that uses optical fibres to locate the exact position of a bullet's impact. Among other interesting functions, this property of location enables a soldier or policemen to carry out health and vital function analysis in a combat situation.

In the present study, optical fibres in textile structures are used to create flexible textile-based displays based on fabrics made of optical fibres and classic yarns.[30-32] The screen matrix is created during weaving, using the texture of the fabric. Integrated into the system is a small electronics interface that controls the LEDs that light groups of fibres. Each group provides light to one given area of the matrix. Specific control of the LEDs then enables various patterns to be displayed in a static or dynamic manner. The basic concept of flexible display is described. It includes the weaving phase, the optical fibre processing procedure that creates the pattern matrices, the electronics interface that controls these matrices and several applications of flexible displays. The two main interesting characteristics of this new flexible device are its very thin size and the fact that it is ultra lightweight. This leads one to believe that such a device could quickly enable innovative solutions for numerous applications.

8.3.3 Optical fibre flexible display (OFFD)

In this section, the process of conceiving and realising an optical fibre flexible display is discussed in detail.

Weaving of optical fibres

Poly(methylmethacrylate) (PMMA) optical fibres present a rigidity and a fragility that are superior to the majority of traditional textile fibre threads and filaments. A good compromise must be obtained between a too significant section diameter synonymous with rigidity, and a too small diameter that induces a low shear resistance and a loss of light intensity. Fibres of diameter of 0.5 mm were retained to make the first prototypes. Tests on fibres with a diameter of 0.25 mm were carried out, but developments in the process of weaving are still required to ensure sufficient fabric resistance in bending.

Weaving is carried out on a traditional 2D loom. The optical fibres can be weaved or placed in a chain, in addition to other kinds of yarns. Therefore, it is theoretically possible to obtain an optical fibre X–Y network. However, this would present several disadvantages:

- The fabric would be extremely rigid and the grid (and thus the resolution) not very dense because of the relatively high radius of curvature of optical fibres.
- Constituting an optical fibre chain is very long and very expensive.
- The resolution would be tiny if one refers to the matrix processing described below.
- It is also possible to note that a 3D-structure in weaving would not bring any advantages.

Thus, the initial plan was to realise a fabric comprising optical fibres for wefts and silk in chain. Other natural, artificial or synthetic yarns could also have been retained to constitute the chain. The choice of yarns for the chain must nevertheless be guided by the aim of achieving good flexibility in the fabric, fine titration and an improved capacity to diffuse and reflect the light emitted by optical fibres for better legibility of information. Moreover, different textile finishings are being tested, either in pasting, or in coating, to guarantee grid stability and to enable optimal light emission intensity and contrast.

Display matrix design

The screen comprises a number of surface units, which can be lit individually. The principle of operation consists in conveying to each 'pixel' a light source emitted from one side of the fabric by one or several PMMA optical fibres with discrete index variation. Each 'pixel' is directly formed on optical fibres while transversely making a spout of light at this precise point on the fabric. This is made possible by a patented process of fibre surface mechanical attack invented by a French company.[33] The process consists of generating microperforations that reach into the heart of the fibre (Fig. 8.4). The remainder of the optical fibre, which did not receive any specific processing, conveys the light without being visible on the surface. The diameter of the fibre section can be as little as 0.25 mm, providing extremely thin and light fabrics.

Communication apparel and optical fibre fabric display

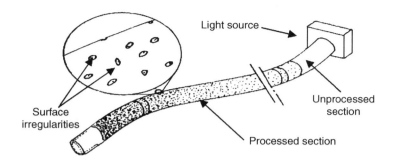

8.4 Optical fibre with multiple point lateral illumination.

Up to now, three existing methods have been available to light ON and OFF static patterns on the fabric (texts, logos and scanned pictures):

- According to the first method, a basic fabric is used. The lighting zone to be processed, comprising optical fibres, is delimited by a stencil key. The picture remains static – with eventual colour changes – but can offer quite a high resolution (Fig. 8.6, details 1 and 2).
- In the second method, the zone to be lit is formed during weaving on a Jacquard loom before being processed. The remaining, inactive fabric is made up of the floating fibres on the back of the fabric.
- In the third method, a two-layer adapted basic-velour fabric is used that makes optical fibres as visible as possible, but with sufficient consistency of fabric structure. Before the weaving process the optical fibres are chemically treated, enabling the specific dynamical lighting zones to be created.[32]

These techniques had to be adapted in order to generate variable information on the same fabric zone, by creating specific weaving armour and an adapted lighting control. The process described below refers to a matrix that makes it possible to display a great deal of basic information, such as texts, logos or other patterns, in a static or dynamic way.[33, 34]

Since the display can only be driven according to columns made of a single optical fibre or groups of fibres (see page 166), lines had to be artificially created. Taking the example of two superimposed patterns to be lightened on the same column, the principle consists of alternating two consecutive weft fibres, one being intended for the first pattern, the other one for the second pattern. Each one is processed on a precise section in order to re-emit light at the place concerned (Fig. 8.5). The principle is the same for three superimposed patterns, taking one fibre out of three for each pattern. By adopting a sufficiently tight weaving, a visual impression is given of full, enlightened zones. Chain wires will be able, according to the material and texture that compose them, to contribute to diffuse the light

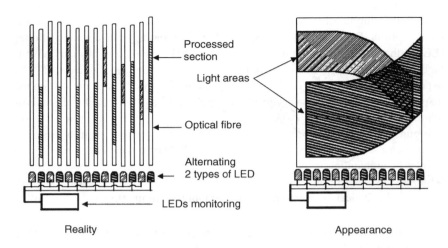

8.5 Two independent light areas superimposed on a single piece of fabric.

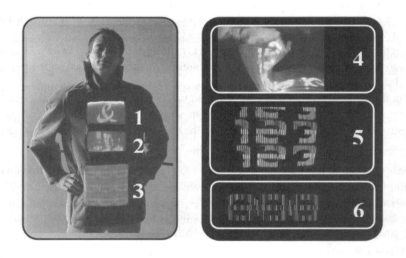

8.6 First prototype at France Telecom Recherche et Développement (FTR&D) using an optical fibre fabric display.

towards the dark zones between enlightened segments. The number of rows to be produced seems limited by the technique, insofar as, on the same unit zone, one quickly obtains more dark zones than enlightened ones – in theory starting from three lines. The appreciation of the definition will then be done according to the size of the screen and the number of pixels in addition to the distance from which people watch this screen.

Various light sources can be used to feed the matrix. The choice mainly depends on both the number of fibres connected to each source and on the level of power consumption. For the first prototypes, high luminous light-emitting diodes (LEDs) that are 3 mm in diameter were used. LED technology has many advantages, as diodes can be easily driven by electronics under low voltages (2–4 V, depending on the colour). Therefore, many 'light effects' can be generated on the display, such as flashing or varying the intensity of the light, providing all kinds of animated 'movies'.

Figure 8.6 shows the very first OFFD, which was carried out in a jacket. It comprises a screen matrix specially designed to display, on one line, three 60 mm × 60 mm alphanumeric characters, each made up of three 'rows' and three 'columns' using 0.5 mm diameter optical fibres and a 7 fibres/cm width density (Fig. 8.6, details 3, 5, 6). Each 'pixel' made up of four fibre segments is controlled by one LED located in the lining of the cloth, on one side of the OFFD. The colour of the 'pixels' is determined by the corresponding LEDs.

On this prototype, it should be noted that OFFD offers another possibility. If, on the one hand, the definition is limited by the number of rows, on the other it is possible to repeat the same line of characters or patterns on the fabric in the direction imposed by optical fibres (Fig. 8.6, detail 5). If repeating the same text can appear useless, the fixed or animated pattern reproduction can be utilised for purely decorative applications, for example, to create a mural tapestry adapting its colours to the clothes worn by the occupants of a room.

The next step in the development of OFFD technology was to increase the density of the optical fibre weave as the purpose was to create more independent 'rows' (see Fig. 8.5) in order to display more than just alphanumeric characters. A second prototype shown in Fig. 8.7 and Fig. 8.8, was designed to match other conventional display matrices. This new display is made of eight rows and eight columns, and runs according to the same principle as for the first prototype (Fig. 8.7). Animated 'movies' are, of course, not represented in this paper, but some static illustrations of patterns that have been made possible are given in Fig. 8.8.

Figure 8.9 shows two prototypes with a display of 8 × 8 pixels that have been realised recently. The first is a bag and the second a communicative jacket. Explanations and several properties of these new prototypes are given below.

Optical fibre screens provide access to simple and animated visual information, such as texts or pictograms. It is possible to download, create or exchange your own visuals via the appropriate internet gateway. The sending of images or text using wireless technology from a computer or a mobile internet terminal to an article of clothing is also envisaged.

The main functions of the new prototypes are:

- to 'be seen', for security, publicity, games or aesthetic purposes
- to show one's affiliation or support for a group

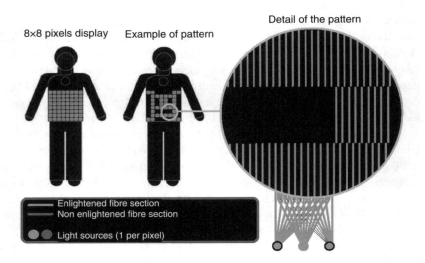

8.7 Second prototype of OFFD at FTR&D.

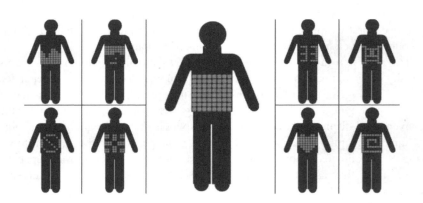

8.8 Examples of patterns on the 8 × 8 matrix.

- to personalise one's clothing according to the latest fashions
- to communicate, or exchange information or to signpost advice.

The interests and advantages of the concept are:

- the ability to create and download animations from a fixed or mobile internet gateway to an article of clothing or a clothing accessory
- to animate a forum online
- to manage information on one's clothing in real time.

Communication apparel and optical fibre fabric display

8.9 Recent prototypes with an 8 × 8 matrix.

8.3.4 Electronics and telecommunication services

The optical fibre matrices developed in the previous section are electronically controlled in order to display different characters and patterns in a static or dynamic way. A keyboard (Fig. 8.10) integrated into the clothing is used to control the display, including several functions: eight preset 'movies' are stored in the electronics memory and can be played in different modes, such as setting up the screen light intensity or interaction with sounds and gestures thanks to specific sensors, and so on. In this prototype, light and small batteries ensure two hours of operation time.

Telecommunication services are shown in Fig. 8.11. The development of this prototype also includes an important software component that allows the user to create his or her own patterns and movies online, using a PC or mobile device.

Because the device is connected to the internet, people can edit the presets and even get new ones already available from a password-protected database.

8.3.5 Flexible display application

Research on the design and development of flexible displays based on optically treated fibres has opened up new frontiers in fields such as smart and communicative clothing, car equipment and home equipment and decoration. The broad range of applications for flexible displays in a variety of segments is summarised in Table 8.1. The table shows a brief overview of the various applications and target populations that could use this technology.

Leisure/business: OFFD can be used as displays for different kinds of mobile phones, personal digital assistants (PDAs), wearable computers and other portable electronic devices. In the fashion industry, different types of fabrics based on

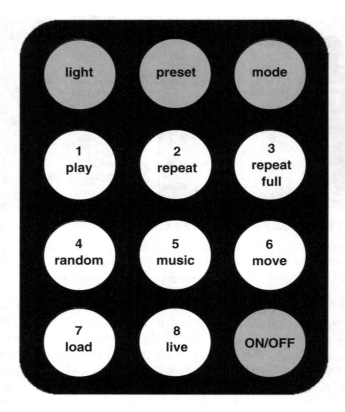

8.10 OFFD remote control.

flexible display technology will open up new horizons for fashion designers and creators.

Public safety: There is enormous potential for firefighting and police applications that display information and warnings on clothes. Therefore, this new functionality increases the personal safety and ability of these people to operate in remote and challenging situations.

Car industry: Many applications of flexible display technology are possible in the car industry. The interior equipment of a car contains many flexible elements that can be used to display relevant information.

House/buildings: Decoration and intelligent houses (buildings) need different flexible support for information, drawings, pictures and lighting. In the case of applications for houses/buildings, the problem of energy supply is easy to resolve.

8.3.6 Future investigations

Existing technologies for weaving and specific processing of optical fibres have been adapted by creating a matrix within the fabric and an electronic control

Communication apparel and optical fibre fabric display 173

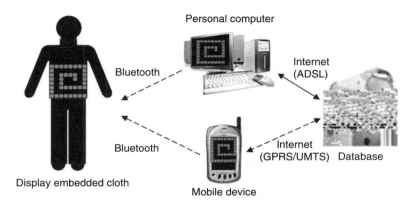

8.11 Telecommunication services. ADSL = asymmetric digital subscriber line, GPRS = general packet radio service, UMTS = universal mobile telecommunication system.

network of LEDs has been developed accordingly to produce an extremely fine, flexible and bright textile display. The structure and the textile materials used suggest a new approach in the field of displays, and more particularly, flexible displays. Generally, textiles have all the basic tools, which are adaptable, to enable the creation of new designs and new apparatuses that will lead to new solutions for specific applications. It is obvious that information is virtually everywhere and that screens and displays have to adopt a multitude of technologies and forms dedicated to targeted applications in public or private places. For this reason, bright optical fibre fabric displays have a significant role to play, particularly in the field of very large flexible displays.

Table 8.1 OFFD applications

Segment	Application type	Target population
Civilian	Leisure/business	Adolescents; adults; businessmen; people using wearable computers, E-mails, fashion clothes...
Public safety	Fire fighting/law enforcement	Firefighters, police
Car industry	External indicators, inside displays	Car makers, designers
House/buildings	Decoration, indication	Architects, decorators...

8.4 Acknowledgements

The authors would like to thank Dubar Warneton,[35] Cédric Brochier Soieries[36] and Audio Images[37] for their contribution to the development of the optical fibre fabric. They also would like to thank Michel Remy, Emeric Mourot and Gemma Ricci for their work and for their continued technical support on the hardware and software aspects of the project at FTR&D.

8.5 References

1. Halary C, 'L'ordinateur vestimentaire en devenir, Seconde peau au clone privé', 2000.
2. Bauer M, Kortuem G and Segall Z, 'Where are you pointing at? A study of remote collaboration in wearable videoconference system', *Proceedings of Third International Symposium on Wearable Computers*, San Francisco, USA, IEEE, 18–19 October 1999.
3. Smailagic A, Siewiorek D, Bass D, Iannucci B, Dahbura A, Eddleston S, Hanson B and Chang E, 'MoCCA: A mobile communication and computing architecture', *Proceedings of Third International Symposium on Wearable Computers*, San Francisco, USA, IEEE, 18–19 October 1999.
4. Vital Signs Monitor, Fitsens, http://www.fitsense.com/, FitSense Technology, 21 Boston Road, PO Box 730, Southborough, MA 01772.
5. Yang J, Yang W, Denecke M and Waibel A, 'Smart sight: A tourist assistant system', *Proceedings of Third International Symposium on Wearable Computers*, San Francisco, USA, IEEE, 18–19 October 1999.
6. Farringdon J, Moore A J, Tilbury N, Church J and Biemond P D, 'Wearable sensor badge and sensor jacket for context awarness', *Proceedings of Third International Symposium on Wearable Computers*, San Francisco, USA, IEEE, 18–19 October 1999.
7. Sangle-capteur de Polar, http://www.randburg.com/fi/polarele.html, Polar Electro Oy Professorintie 5, FIN-90440 Kempele, Finland.
8. Hong Z Tan and Pentland A, 'Tactual displays for wearable computing', *Proceedings of First International Symposium on Wearable Computers*, The Medialab Massachusetts Institute of Technology, Boston, USA, IEEE, 18–19 October 1997.
9. Thomas B, Grimmer K, Makovec D, Zucco J and Gunther B, 'Determination of placement of a body-attached mouse as a pointing input device for wearable computers', *Proceedings of Third International Symposium on Wearable Computers*, San Francisco, USA, IEEE, 18–19 October 1999.
10. Vardy A, Robinson J and Cheng L-T, 'The WristCam as input device', *Proceedings of Third International Symposium on Wearable Computers*, San Francisco, USA, IEEE, 18–19 October 1999.
11. Rischpater R, Project: the Spyglass: an interface for I-wear, *I-Wear Consortium Meeting*, Brussels, Belgium, Starlab, 6–7 June 2000, 30–35.
12. Cleveland G and McNinch L, 'Force XXI land warrior: implementing spoken commands for soldier wearable systems', *Proceedings of Third International Symposium on Wearable Computers*, IEEE, 18–19 October 1999.
13. Smailagic A, Siewiorek D, Martin R and Reilly D, 'CMU wearable computers for real-time speech translation', *Proceedings of Third International Symposium on*

Wearable Computers, San Francisco, USA, IEEE, 18–19 October 1999.
14. Kangchun Perng J, Fisher B, Hollar S and Pister K S J, 'Acceleration sensing glove', *Proceedings of Third International Symposium on Wearable Computers*, San Francisco, USA, IEEE, 18–19 October 1999.
15. Pratt V R, 'Thumbcode: A Device-independent Digital Sign Language', Stanford University Report, USA, July 1998, http://boole.stanford.edu/thumbcode/
16. Mills G, Slaten L, Broughton R, Malone K and Taylor D, 'Photoadaptive fibres for textile materials', *National Textile Centre Research Report* M98-A10, Volume 8, November 2000.
17. Gregory R V, Samuels R J and Hanks T, 'Chameleon fibres: dynamic color change from tunable molecular and oligomeric devices', *National Textile Centre Research Report* M98-C1, Volume 8, November 2000.
18. Gregory R V, Kimbrell W C and Kuhn H H, *Synth. Met.*, 1989, **28**(1&2), c823.
19. Gregory R, *Handbook of Conductive Polymers*, Second edn, Skotheim T R, Elsenbaumer R and Reynolds J R (eds), Portland, OR, Marcel Dekker, 1997.
20. Visson Israel Ltd,1 Bezalel Street, Ramat Gan 52521, Israel, www.electronicsweekly.com/Article24123.htm
21. Kim Y K, Warner S, Lewis A and Kumar S, 'Nanocomposite fibres', *National Textile Centre Research Report* M00-D08, Volume 8, November 2000.
22. Du C W, Tao X M., Tam Y L and Choy C L, 'Fundamentals and applications of optical fibre Bragg grating sensors to textile composites', *J. Composite Struct.*, 1998, **42**(3), 217–230.
23. Tang L Q, Tao X M, Du W C and Choy C L, 'Reliability of fibre Bragg grating sensors in textile composites', *J. Composite Interfaces*, 1998, **5**(5), 421–435.
24. Du W C, Tao X M and Tam H Y, 'Fibre Bragg grating cavity sensor for simultaneous measurement of strain and temperature', *IEEE Photonics Technol. Lett.*, 1999, **11**(1), 105–107.
25. Tao X M, 'Integration of fibre optic sensors in smart textile composites – design and fabrication', *J. Text. Inst.*, 2000, **91**(Part 1, no. 3), 448–459.
26. Jackson D A, Ribeiro A B L, Reekie L, Archambault J L and Russell P St, 'Simultaneous interrogation of fibre optic grating sensors', *Proceedings OFS'9*, Florence, Italy, 1993.
27. Kersey A D, Davis M A, and Morey W W, 'Quasi-distributed Bragg-grating fibre-laser sensor', *Proceedings OFS'9*, Florence Italy, 1993, postdeadline paper PD-5.
28. Xu M G, Reekie L, Chow Y T and Dakin J P, 'Optical in-fibre grating high pressure sensor', *Electron. Lett.*, 1993, **29**(4), 398–399.
29. Jayaraman S, 'The first fully computerized clothing: A higher quality of life through technology', *Second International Avantex Symposium*, Frankfurt, Germany, 2002.
30. Deflin E and Koncar V, 'For communicating clothing: The flexible display of glass fibre fabrics is reality', *Second International Avantex Symposium*, Frankfurt, Germany, 2002.
31. Deflin E, Weill A, Koncar V and Vinchon H, 'Bright optical fiber fabric – a new flexible display', *The Sixth Asian Textile Conference*, 22–24 August 2001, Hong Kong, CD ROM Proceedings.
32. Veyet F and Koncar V, 'Innovation Textile: Les parapluies intègrent les écrans lumineux', *Rapport de Projet de Fin d'Etudes*, ENSAIT, Roubaix, France, June, 2002.
33. Bernasson A and Vergne H, *Optical Fiber with Multiple Point Lateral Illumination*, International Patent no. PCT/FR94/01475, 1998.

34. Deflin E, Weill A, Ricci G and Bonfiglio J, *Dispositif Lumineux Comprenant une Multiplicité de Fibres optiques à Segments Lumineux*, Brevet n°0102623 déposé en France par France Télécom le 27/02/01.
35. Vinchon H, Dubar Warneton, 136, rue Jules Guesde B.P.189, 59391 WATTRELOS CEDEX – FRANCE, http://www.dubar-warneton.com
36. Cedric B, Soieries 33, rue Romarin 69001 LYON, http://www.cedricbrochiersoieries.com
37. Bernasson A, Sarl Audio Images, Parc Ind. du Maréchat, 2 rue A. Einstein, 63200 RIOM – France, http://www.excel-ray.com

9
Wearable computing systems – electronic textiles

TÜNDE KIRSTEIN, DIDIER COTTET,
JANUSZ GRZYB and GERHARD TRÖSTER
Swiss Federal Institute of Technology Zurich, Switzerland

9.1 Introduction

The vision behind the idea of wearable computing systems describes future electronic systems as an integral part of our everyday clothing, serving us as intelligent personal assistants. A wearable computer is always on, does not hinder the user's activities, has easy-to-use interfaces, is aware of the user's situation and provides support, e.g. by displaying relevant information, monitoring health parameters and augmenting the user's view of reality. The first wearable computers were developed for navigation and maintenance tasks (e.g. aircraft inspection[1]) and for military applications. The Defense Advanced Research Projects Agency (DARPA) supports wearable computing research in the United States in order to achieve a breakthrough in developments in body-worn computational resources for soldiers. In the meantime, wearable computing is predicted to have a future in daily life acting as a general-purpose system rather than a single-purpose system. Mann[2] regards wearable computers as a 'second brain' and their sensory modalities as additional senses augmenting human intellect. In this way, wearable computers can contribute to the vision of an 'ambient intelligence', where intelligent devices are integrated into the everyday environment and provide a multitude of services for everyone. Starner[3] describes the possibilities offered by wearable systems and discusses challenges in wearable research.

Figure 9.1 shows the possible systemisation of wearable computers. The first level consists of the components of a wearable system that provide several functions:

- sensor unit: registration of biometric and environmental data and of user commands
- network unit: transmission of data within the wearable computer and to external networks
- processing unit: calculating, analysing and storing data

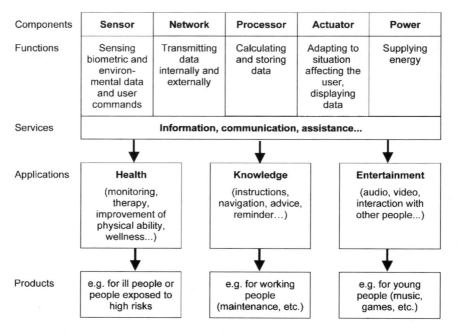

9.1 Systemisation of wearable electronic systems.

- power unit: supplying energy
- actuator unit: adapting to situations, creating an effect on the user, displaying data.

On the system level, several of these functions are combined to form services. Providing information, communication or assistance are possible services. Because mobility is now a fundamental aspect of many services and devices, there are an almost unlimited number of application ideas, e.g. in the fields of health, knowledge and entertainment. Health applications include the monitoring of ill or high-risk persons, people exposed to extreme conditions or people engaged in sports, but also therapy and improvement of physical abilities. Examples of knowledge applications are navigation or instructions for work. There are a large number of applications also in the area of entertainment (e.g. audio and video). The last level shows possible target groups for wearable electronic products.

9.2 Why is clothing an ideal place for intelligent systems?

Clothing is an important and special part of our environment, as it is personal, comfortable, close to the body and used almost anywhere and anytime. Nowadays, clothing has more functions than just climatic protection and good looks. Zhang and Tao[4-7] give an overview of clothing that is considered to be smart. For example, shape memory textiles and phase change materials adapt their properties depending on the temperature and have found application in climate-regulating garments. But clothing is still far from taking full advantage of the potential of information technology services. If clothing had intelligent features, it could serve us in a very unobtrusive and natural way. Clothing could 'enhance our capabilities without requiring any conscious thought or effort'.[8] Being close to the body, clothing enables the intimate interaction of man and machine. This interaction is necessary for any kind of computer intelligence to be used for context recognition or as intuitive interfaces. The WearNET is an example of a sensor system attached to the body that can provide a wearable computer with a wide range of context information.[9]

However, most of the wearable computers on the market are not integrated into clothing. They still consist of bulky and rigid boxes and are portable machines rather than a comfortable part of the clothing (e.g. Xybernaut[10]). Gemperle et al.[11] describe how the shape and placement of such devices can be improved in terms of wearability. The WearArm that has been developed by ETH and Massachussetts Institute of Technology (MIT) is a wearable computer that combines complex functionality and high speed with advanced wearability.[12] This computer consists of miniaturised modules connected using flexible substrates and equipped with advanced, context-sensitive power-management features.

The next step towards real wearability is the integration into clothing. There are two methods of integration:

- the miniaturisation of electronic components and their attachment to textiles
- the development of textiles with electronic functions (electronic textiles).

Clothing for the arctic environment,[13] the Philips jacket[14] and the 'Lifeshirt'[15] are some examples of the first method. In these products the textiles simply serve as a carrier of conventional cables, special connectors and miniaturised electronic devices. Humans prefer to wear textiles, as they are flexible, soft, lightweight, breathable, robust and washable. Thus, the idea emerged of developing fibres and fabrics that can be used for electronic functions. A suitable definition considers electronic textiles as materials possessing both electronic functionality and textile characteristics.

9.3 Electronic textiles

In order to describe electronic textiles it is necessary to define the term 'textile'. Materials are considered to be a textile when they consist of drapeable structures that can be processed on textile machinery. Textiles are usually made of fine and flexible fibres and threads that have a high length/diameter ratio but textiles can also contain membranes and foils. Ready-made textile products include ropes, ribbons, fabrics and also three-dimensional products such as clothing.

'Electronic' means that a system is able to exchange and process information. If textiles had the ability to record, analyse, store, send and display data, a new dimension of intelligent high-tech clothing could be reached. However, there are some general difficulties in creating electronic textiles for clothing. Electrical functions have to be embedded in textiles in such a way that the flexibility and comfort of the fabrics are retained. Fibres and fabrics have to meet special requirements concerning not only conductivity but also processability and wearability:

- The fibres have to be able to withstand handling that is typical for textiles, for example weaving, washing and wrinkling, without damaging functionality.
- Fibres that are used for clothing have to be fine and somewhat elastic in order to be comfortable to wear.
- Fabrics need to have a low mechanical resistance to bending and shearing so that they can be easily deformed and draped. The closer the textiles are to the body, the more flexible and lightweight they have to be.

These demands are partly inconsistent with the materials and geometries that are needed for conducting data and power. There are two possible kinds of conductivity: electrical or optical conductivity. Optical fibres have some advantages, as there can be no shorts, no corrosion and no parasitic field effects. But electrical conductors are easier to handle in textile fabrication processes. In addition, the costs are higher for optical fibres owing to expensive light sources and transceivers. Most plastic optical fibres are stiff, allow a limited bending radius and are difficult to weave or knit. Metal, carbon and conductive polymers are also quite rigid and brittle materials. They are heavier than most textile fibres, making homogeneous blends difficult to produce. Nevertheless, textile technologies have been developed to manufacture processable fibres and yarns from these materials.[16,17] Conductive fabrics already have found applications, especially in the field of electromagnetic interference (EMI) shielding and static dissipation. Methods of creating electrically conductive fibres are:

- filling synthetic fibres with carbon or metal particles
- coating fibres with conductive polymers or metal
- using continuous or short fibres that are completely made of conductive material.

These fibres can be woven, knitted or embroidered into fabrics. Conductive ink technology offers another alternative in the development of electronic textiles. The benefits offered by digital printing techniques have prompted many conductive ink developers to experiment with printing onto textile substrates (e.g. Colortronics[18]).

9.3.1 Textile networks

Electrical networks are responsible for data and power transfer. On-body communications in wearable systems can be wired or wireless. A textile network means that fabrics are used to replace conventional wires, whole circuit boards or antennas. The first efforts to use conductive textiles for electrical circuitry were made at the MIT Media Lab in the E-broidery project.[19] Conductor lines were realised by embroidering metal fibres or weaving silk threads that were wrapped in thin copper foil. Gorlick's 'electric suspenders'[20] contain stainless steel conductors for power and data buses. The main drawback of these prototypes is the need for protection against shorting and corrosion, as the conductive fibres are not insulated. Jayaraman and co-workers[21] developed a garment that included electrically conductive fibres and plastic optic fibres to transfer information from sensors to processing units. In this so-called 'wearable motherboard', each electrical fibre (e.g. stainless steel, copper or doped nylon fibre) is insulated with a polyvinylchloride (PVC) or polyethylene coating. At ETH Zurich extensive studies were carried out to measure and model the high frequency properties of conductive fabrics (see Section 9.4). With the developed model it is possible to simulate and optimise communication networks in textiles.

In the area of communication architecture there are approaches to investigate networks of distributed computing elements embedded into fabrics. Marculescu *et al.*[22] describe computational models of large-area information systems that adapt to changing conditions and reconfigure on-the-fly to achieve better functionality. The reconfigurable fabric of UCLA[23] contains organic electronic devices connected to flexible polymeric substrates. These devices have reconfigurable software so that the network can configure itself based on initially available resources and ongoing damage monitoring.

Kallmayer and Griese[24] realised transponder antennas by weaving conductive fibres into multilayer fabrics. Hum[25] used several short-range fabric antenna coils to create a wireless communication infrastructure between various positions on clothing and between different pieces of clothing (FAN = fabric area network, multiple radio frequency identification technology (RFID) links). These antennas can be used to communicate with transponder chips that are embedded, for example in personal items. Massey[26] integrated mobile phone antennas in clothing by using the relatively large surface area of the fabric to include a ground plane and an antenna element made of copper-plated textiles.

9.3.2 Textile sensors and actuators

The next substantial progress will be to use textiles not only for transmitting signals but also for transforming them. Transformation of signals can happen in two ways:

- sensor = transformation of physical phenomena into processable electrical signals
- actuator = transformation of electrical signals into physical phenomena.

Textile technologies have been developed to create fibres and fabrics with a significant and reproducible change of properties caused by defined environmental influences. Sensors can be used to measure biometric or environmental data, but also to act as an input interface. Actuators can adapt themselves to a situation, affect the human body or serve as a display.

'SOFTswitch'[27] is one example of a textile pressure sensor. It is made of conductive fabrics with a thin layer of elastoresistive composite that reduces its resistance when it is compressed. The 'sensory fabric'[28] also consists of two conductive fabric layers separated by a nonconductive layer. This product has an even simpler construction because the nonconductive layer is formed by a mesh so that pressure can create a contact in the holes of the mesh. This pressure-sensitive fabric can be used for such items as soft keypads. The 'sensor jacket'[29] measures stretch from resistive changes in knitted strips. Gorix[30] is a woven textile formed from pre-oxidised carbonised polymers that can be used, not only for resistive heating, but also as a temperature or pressure sensor.

In the 'Wearable Motherboard',[21] plastic optical fibres detect damage (broken paths) in the fabric and can give information about the location of bullet penetration. Tao et al.[31] describe the use of fibre optic sensors inside fabric-reinforced structures that monitor mechanical, acoustic, electric, magnetic and thermal perturbations. Fibre Bragg-grating sensors are manufactured by modifying the core in a single-mode optic fibre to detect the wavelength shift induced by strain or temperature change. Sensing can be distributed along the length of the fibre and it is possible to measure different parameters simultaneously. The possibilities of integrating fibre optic sensors in clothing seem promising. El-Sherif et al.[32] have embedded fibre optic sensors into soldiers' uniforms. These sensors can detect chemical, biological and thermal hazards. The optical fibres have a sensitive cladding that can change the light propagation characteristic of the fibre.

During the last few years polymers have emerged that respond to electrical stimulation with property changes (e.g. shape, size, colour change). These polymers are called 'electroactive polymers' and can be used as sensors or actuators.[33] Gregory et al.[34] developed fibres that can quickly change their colour or optical transparency by the application of an electrical or magnetic field. De Rossi et al.[35] describe fabrics coated with a thin layer of conducting polymer that possess strain and temperature sensing properties and also actuating properties. These fabrics contain fibre bundles that contract and relax under electrical control and can be used as a tactile output interface (e.g. in a sensor glove).

France Telecom[36] has developed a display made of optical fibres woven into a fabric. However, owing to the mechanical limitations of the optical fibres, the number of pixels is just 64 and the fibre diameter is 0.5 mm. Another technology that could possibly be adapted to textiles is the fabrication of organic light-emitting diodes (LEDs). Processes have already been developed to create such displays on flexible substrates but not yet on textiles. Textile displays also can be realised using conductive fibres covered with a fine layer of an electroluminescent material.

9.3.3 Textile processors

Processing includes arithmetic operations and the storage of data. Transistors, diodes and other non-linear devices are needed for these functions. Manufacturing technologies have already been developed to create organic devices ('all-polymer' transistors and batteries). Such devices, as well as thin silicon integrated circuits (ICs) can be fabricated on flexible polymer substrates.[24] Such flexible chips can be attached to textiles but they are not textiles themselves. Preliminary efforts are being made to create textile fibres from electroactive polymers that can act as transistors. One of the main challenges is to improve the stability of such conductive fibres. Apart from the short lifetime, the slow switching speed is another limitation of performance. A company[37] has created threads with transistor functionality and has been able to make them more stable so that they can perform millions of switching operations.

9.3.4 Textile power supply

As the power supply is usually the biggest and heaviest part in the wearable computers of today, there are several approaches to decreasing power consumption by power management or power performance trade-offs and to developing novel power supply technologies (e.g. lithium polymer batteries, micro fuel cells[38]). The alternative to batteries is to use different sources of energy available on the body which can be transformed into electrical energy, e.g. sunlight, body temperature, body motion.[39] Some efforts have been made to embed miniaturised or flexible energy supplies into textiles, but few to create a textile power supply. Infineon[40] presented miniaturised silicon thermoelectric generators attached to clothing. Using the difference in temperature between the outside and the inside of clothing, these thermogenerators produce an output power of a few microwatts per square centimetre. Thin film solar cells can be made on flexible surfaces such as plastics. Baps et al.[41] adapted the technology for the flexible solar cell to fibre form. The efficiency of these alternative energy sources needs to be improved. Creating components that are wirelessly powered by an electric field in the environment is another interesting approach. For example, sensors or devices could 'wake up' only when the user is close to an electrical field.

9.3.5 Potential of electronic textiles

This overview shows that electronics cannot only be attached to textiles but also realised in the form of textile structures. Of course, the performance and costs of such textile structures cannot be compared with those of conventional computer technology. There are strong limitations concerning mass production and reliability. But apart from the difficulties though, there is also potential for the use of electronic textiles. Textiles offer new and fascinating possibilities in creating information systems. The geometric and mechanical properties of textiles (large flexible area) differ strongly from conventional electronics and can create new computer designs and architectures.[22,23]

In the future, it could become quite difficult to separate clearly electronic textiles from the aforementioned method of miniaturisation and attachment, because computers could be miniaturised until they are the size of molecules. In this case, 'attachment' to fibres or fabrics would also result in electronic textiles.

9.4 Electrical characterisation of textile networks

This section presents the results of an interdisciplinary research work of textile and electrical engineers at ETH Zurich. We wanted to characterise communication networks in fabrics. We developed methods for measuring and modelling the high frequency properties of textile transmission lines in order to discover the limits and potential of textile-based communications. Our aim is to replace conventional wires and even high-performance circuit boards with textile fabrics. Therefore, we applied methods of microwave technology. That means that wires are not only characterised by their ohmic resistance but by wave effects depending on the line geometries and the surrounding material. Therefore, we also had to consider the geometric structures that are created in the textile fabrication processes.

9.4.1 Textile geometry

The geometry of textile materials is characterised by a hierarchical structure: bundles of fibres are twisted to create yarns; yarns are woven or knitted to create fabrics. Fabrics for signal transmission in wearable computers have to meet requirements concerning processability and wearability (see Section 9.3). It is necessary to have individually addressable conductors that are insulated to prevent shorts.

For our experiments, we fabricated yarns that contain insulated metal filaments and fulfil the aforementioned requirements. The characterisation methods that we developed can be applied to all kinds of conductive textiles. We chose woven fabrics with a plain weave in our experiments because this construction represents the most elementary and simple textile structure. In addition, this kind of material can provide a tight mesh of individually addressable wires that can be used for basic transmission lines as well as for whole circuits.

The fabrics that were examined contain polyester (PES) yarns that are twisted with a conductive fibre (copper). The conductive fibres have a diameter of 40 μm and are insulated with a polyesterimide coating. PES yarns with two different thicknesses (167×10^{-4} g m^{-1} resp. 334×10^{-4} g m^{-1}) have been used to create six different fabric types (Table 9.1). All fabrics have about 20 threads per cm in both directions but their densities differ according to the PES fineness used. Two of the fabrics have conductive fibres in both directions (warp and weft), two have

Table 9.1 Textiles used in the experiments

Woven fabrics:	Yarn types	Description
Fabric 1	Yarn A and B	Low density with Cu in both directions (*XY*)
Fabric 2	Yarn A and B	Low density with Cu in one direction (*X*)
Fabric 3	Yarn A and B	Low density without Cu
Fabric 4	Yarn C and D	High density with Cu in both directions (*XY*)
Fabric 5	Yarn C and D	High density with Cu in one direction (*X*)
Fabric 6	Yarn C and D	High density without Cu

Note: Yarn A: PES yarn 167×10^{-4} g m^{-1} + Cu filament; Yarn B: PES yarn 167×10^{-4} g m^{-1}; Yarn C: PES yarn 334×10^{-4} g m^{-1} + Cu filament; Yarn D: PES yarn 334×10^{-4} g m^{-1}.

conductive fibres in just one direction (warp) and two are without conductors. Figure 9.2 shows fabric 4.

Taking a closer look at textile geometry one can observe that fibres follow a helical path within the yarn. The helical path of the metal fibres can be seen in Fig. 9.2. When the yarns are woven into a fabric they are periodically crimped (Fig. 9.3). This means that the length of the conductive fibres is greater than the length of the fabric. There are several irregularities concerning the location of the fibres within the yarn as well as concerning the location of the yarns within the fabric. These variations are caused by the deformability of the textile material and the degrees of freedom in the manufacturing processes. At the level of fibres and yarns there are variations in diameters and densities (along the thread but also from thread to thread). At the level of fabrics, the distance between yarns varies (Table 9.2). As textile materials have a viscoelastic behaviour, inner tensions relieve over time and the geometry may change (especially as a result of washing).

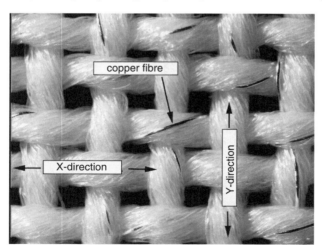

9.2 Woven fabric with conductive fibres.

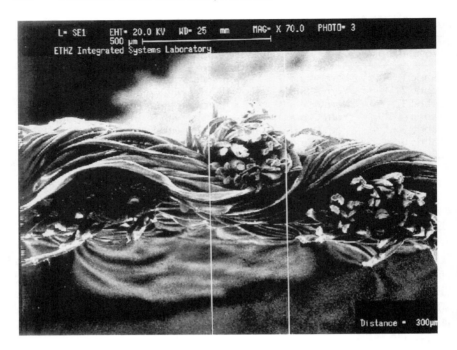

9.3 Fabric cross-section.

Table 9.2 Examples of geometric variations in the textiles used in the experiments

Fabric type	Dimensions (μm)	Variations (μm)
Fabric 1 (low density)	$a = 891$ $d = 228$	$\sigma = 32.9$ $\sigma = 25.3$
Fabric 4 (high density)	$a = 876$ $d = 334$	$\sigma = 25.0$ $\sigma = 28.0$

Note: d = yarn diameter, a = yarn distance.

9.4.2 Electrical characterisation

In order to evaluate the performance and limits of textile transmission lines we extracted the electrical parameters with time and frequency domain analysis and developed a theoretical model that describes signal transmission in textiles.

Material properties

The direct current (DC) resistance of a single metal fibre was 0.15 Ω cm^{-1} for yarn A and 0.17 Ω cm^{-1} for yarn C. These measured DC resistances and the actual diameter of the metal fibres allowed the effective conductor length to be calculated in comparison with the textile length. For the thinner yarns (type A) the conductor was about 7.5% longer than the corresponding textile, with a tolerance of 0.5%. For the thicker yarn (type C) where the conductive fibre ran a larger helical path, this difference increased to about 25.5% with a tolerance of 2.0%.

The dielectric permittivity ε_r of the mixed PES–air textile structure was extracted by means of parallel plate capacitors of known dimensions and using the yarn types B and D (without conductive fibres). The results obtained range from ε_r = 1.4 to 1.6. This inaccuracy was due to the fact that the measured permittivity strongly depends on the ratio of polyester and air. Introducing conductive fibres would also have affected the total permittivity, as the polyesterimide used for the isolation coating showed an $\varepsilon_r > 3$.

Transmission line configuration

To minimise parasitic coupling at high frequencies the ground line should be close to the signal line. In conventional circuit boards a whole ground layer is often used, but creating such a construction in textile fabrics would have several disadvantages. We decided to take conductive fibres in the warp direction (X-direction) as signal lines and the neighbouring conductive fibres on each side as ground lines. These configurations are similar to the conventional coplanar waveguides (CPW) on printed wire boards. The transmission line configurations differ by the number of signal fibres S or ground fibres G. The space between the ground and the signal line is given by the textile construction and cannot be modified. Any attempt to skip conductive fibres to increase the space would yield floating lines evoking undesired parasitic coupling effects. A list of all investigated configurations is given in Fig. 9.4.

9.4 Transmission line configurations.

188　Wearable electronics and photonics

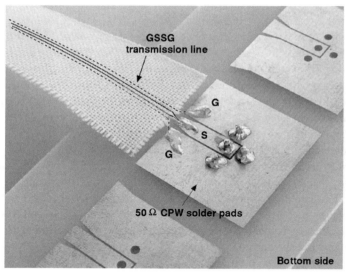

9.5 Textile solder pads as a 50 Ω coplanar waveguide.

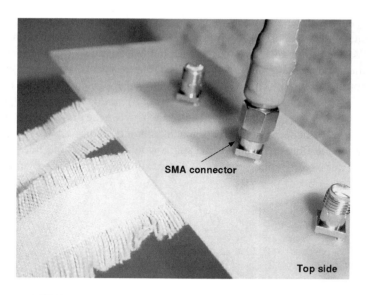

9.6 SMA connectors.

Impedance measurement

We investigated the characteristic impedance of the textile transmission lines. As we expected the textile geometric variations to influence on the impedance, we

Wearable computing systems – electronic textiles 189

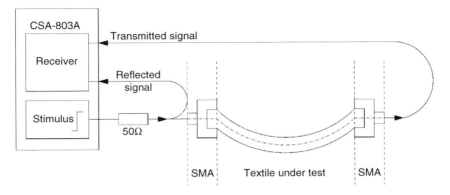

9.7 Time domain reflectrometry (TDR) measurement setup.

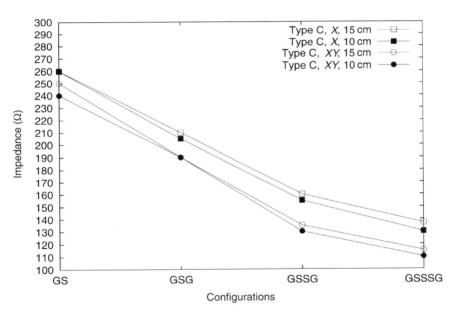

9.8 Measured impedance of different transmission line configurations.

measured the signal reflections along the transmission line with time domain reflectometry. We had to develop suitable connectors to measure the textile lines. FR4 laminate-based interposers with patterned 50 Ω CPW solder pads on one side and surface mount assembly (SMA) connectors on the other side allowed the textile samples to be reliably connected to the measurement equipment (Fig. 9.5 and Fig. 9.6). The block diagram of the measurement setup is depicted in Fig. 9.7.

Figure 9.8 shows typical measured line impedances for the investigated transmission lines. Results for the same yarn and fabric types are connected by lines to

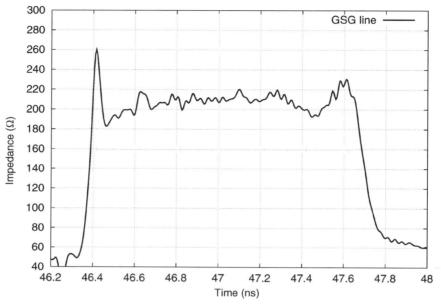

9.9 Impedance profile of 15 cm transmission line.

illustrate the relationship between configuration and line impedance. The *XY* fabrics (metal fibres in both directions) have lower impedances owing to a higher capacitance caused by coupling effects to the floating lines in the *Y* direction. The textiles with conductors only in the *X* direction have a lower capacitance and inductance and therefore provide faster signal propagation than the textiles with metal in the *XY* direction.

The results of the four configurations are comparable to coplanar waveguides on printed circuit boards (PCBs): increasing the signal line width by adding more parallel conductive fibres decreases the line impedance. Using fabrics with smaller distances between the threads (that means between signal and ground lines) would enable a lower line impedance, but 50 Ω lines seem difficult to achieve. The measurements show impedance variations along the textile signal lines (see Fig. 9.9). These variations are caused by geometric irregularities in the fabrics. We investigated the effects of these variations on the predictability of the line impedance.

Impedance simulation

In order to predict the impedance of different fabrics and line configurations we modelled the textile transmission lines with a 2.5 dimensional electromagnetic field solver (Sonnet EM Suite 7.0). We wanted to get a better understanding of how textile fabrication tolerances affect the line impedance. To simplify the modelling

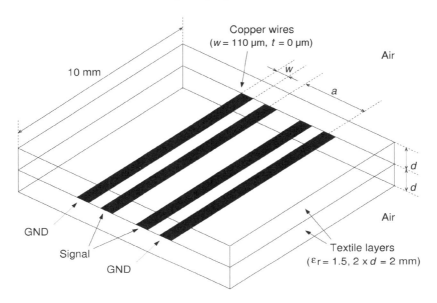

9.10 Textile model for Sonnet showing a GSSG configuration. a = distance between copper wires.

and to reduce the computation time, the structure of the woven fabric was regarded as a homogeneous material with an equivalent dielectric permittivity as previously measured on page 187. In effect the textile model consists of two dielectric layers with a permittivity $\varepsilon_r = 1.5$ and a thickness of 2 mm (see Fig. 9.10). The conductive fibres are modelled as planar strips between the two textile layers. To compensate for the helical shape of the conductor within the yarn, the width of the conductive stripes is averaged to $w = d/\pi \approx 110$ µm, where d is the yarn diameter.

Table 9.3 shows simulated impedances for fabric type C and demonstrates the impedance variations caused by the textile process tolerances. Z_{min} and Z_{max} have been calculated with the 'worst case' textile geometry. With the developed model it is possible to predetermine the textile line impedances and the achievable tolerances.

Table 9.3 Simulation results of textile line impedances with regard to textile process tolerances

Yarn type	Configuration	Z_{nom} [Ω]	Z_{min} [Ω]	Z_{max} [Ω]
C	GS	263	247	278
C	GSG	190	173	200
C	GSSG	163	157	172

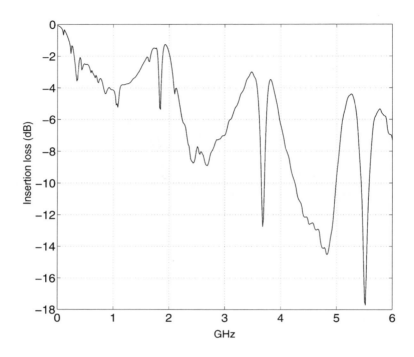

9.11 Measured insertion loss of the 5 cm GSSSG line (fabric 5).

Frequency characterisation

To investigate the frequency characteristics of textile transmission lines we measured the transmission properties with a vector network analyser (VNA) up to 6 GHz. The textiles were connected with the network analyser ports by means of the same FR4 interposers proposed earlier.

As mentioned before, the textile lines are not reflectionless, but stochastically change their impedance value. The variations of wire lengths and distances are able to make the phases unequal and excite parasitic waves ('odd modes'). As the ground wires and signal wires are shorted at the beginning and end of the transmission line, these odd modes have an effect when the line length is a multiple of the half wavelength of the signal. In Fig. 9.11 one can clearly observe some deep minima in the line transmission even down to –18 dB as a result of these parasitic resonances.

We extracted the attenuation constants for the different configurations. The extracted values show that in the lower frequency range, the coupling to the odd modes is weak and single mode propagation can be reasonably justified. One can observe that the extracted attenuation, even in this frequency range, shows non-monotonical frequency behaviour, which is not typical for uniform transmission

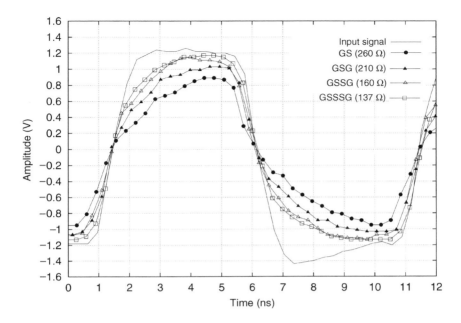

9.12 100 MHz clock signals measured through four different 20 cm-long textile transmission lines.

lines. This is the effect of non-uniform characteristic impedance profiles along the lines caused by large geometric tolerances. The frequency behaviour of the tested configurations shows a similar dependence with the same maximum attenuation of 0.05–0.1 dB cm^{-1} in the frequency range of single mode propagation.

We can arrive at the very important conclusion that the insertion loss of the textile lines is not determined by the dielectric and ohmic losses, but by the reflections along the line in the lower frequency range and coupling to the parasitic modes at higher frequencies above the half wavelength. Although the textile wires feature high conductivity, this plays a minor role in determining the loss factor of the lines. The *XY* configurations show slightly lower losses in the single mode propagation range and weaker but more irregular coupling to odd modes. This is the effect of the orthogonal wires, which are able to destroy the constructive resonances of the odd modes to some extent.

Based on different measured line configurations we can conclude that the longest possible line length is equal to the half wavelength of the lines at the maximal desired frequency of usage. This allows the lines to be 10 cm long for maximal frequencies of approximately 1.2 GHz and 1 GHz for the *X* and *XY* configurations, respectively. For 100 MHz signals the allowable line lengths are tenfold and are in the range of 100 cm.

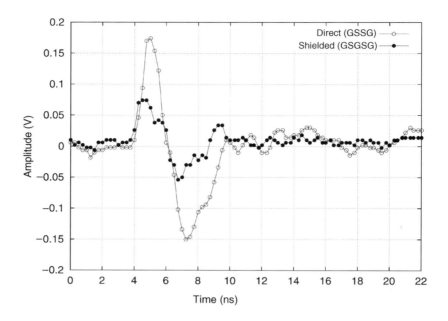

9.13 Far-end crosstalk measured on 20 cm matched load lines with and without shielding.

Signal integrity and crosstalk

Figure 9.12 shows a 100 MHz clock signal measured at the end of 20 cm textile transmission lines in different configurations, and demonstrates the signal integrity.

Figure 9.13 presents the results of the far-end crosstalk for the two neighbouring lines in GS and GSG configurations. The measurements were performed on two 20 cm-long lines terminated in a matched load. The amplitude of the aggressor signal was 2.5 V with a rise time of 6 ns. The ground fibre between the neighbouring lines in the GSG configuration, acting as a shield, was able to reduce the crosstalk from 7.2% in GS configuration down to 2.8%.

9.5 Conclusions

We have presented, for the first time, an extensive characterisation of textile transmission lines for use in wearable computing applications. The proposed textiles are fabrics with conductive fibres in one or both directions and with different yarn fineness. This variety of fabrics opens a wide range of possible transmission line topologies, allowing a configuration to be found that fits the target application.

The FR4 interposer with coplanar solder pads and SMA connectors allowed the textile samples to be reliably connected to the measurement equipment. The TDR measurements showed that the achievable characteristic impedances lie between 120 Ω and 320 Ω. To study the influence of fabrication tolerances, the textiles were modelled with Sonnet, an electromagnetic field simulation tool. The simulation results showed that, with the given geometry variations, an accuracy of ±5% to ±10% for the characteristic impedances is achievable.

High frequency network analyser measurements were performed up to 6 GHz. The extracted frequency characteristics revealed that the dielectric and ohmic losses do not determine the line insertion loss. The line insertion loss is mainly influenced by non-uniform impedance profiles along the lines up to the half wavelength and by coupling to parasitic modes above this frequency point. This results in cut-off frequencies of 1.2 and 1 GHz for 10 cm-long lines in *X* and *XY* configurations, respectively. Good signal transmission for a 100 MHz clock signal was proved through 20 cm textile lines. Experiments also showed that grounded conductor lines between two neighbouring signal lines reduced crosstalk from 7.2 to 2.8%.

The final conclusion of this work is that conductive textiles provide much more than EMI shielding and power supply. Transmission lines with controlled characteristic impedance and high signal integrity up to several 100 MHz enable new options in interconnection for wearable computers.

9.6   Future challenges

To advance from electronic textiles to electronic clothing, research has to be carried out in the following areas:

- clothing technology for manufacturing
- testing under wearing conditions and washing/cleaning treatments
- investigation of reliability.

Electronic textiles have to maintain their functionality through repeated wear and washing cycles. They must not be damaged by constant motion and stress from body movements, perspiration and body heat. At ETH Zurich wearing stress is investigated in order to develop a 'body map' showing the stresses on different parts of the clothing during wear processes. Furthermore, the acceptability of these garments will depend on how comfortable they are. The garments have to support the wearer's thermoregulation and should also be friendly to the skin. The electromagnetic fields emitted by electronic systems in clothing have to be investigated and possible ways have to be found to reduce the effects on the environment and the user. Last but not least, clothing concepts have to be developed. Depending on the application, electronic functionality can be fully integrated or a modular approach can be chosen, where clothing provides a kind of 'platform' for several possible modules. In order to achieve these objectives, close interdisciplinary cooperation in the fields of electrical, textile and clothing technology is necessary.

9.7 References

1. Smailagic A and Siewiorek P, 'Modalities of interaction with CMU wearable computers', *IEEE Personal Commun.*, 1996, **3**(1), February, 14–25.
2. Mann S, 'Wearable computing: toward humanistic intelligence', *IEEE Intelligent Systems*, 2001, May/June, 10–15.
3. Starner T, 'The challenges of wearable computing', *IEEE Micro*, 2001, **21**(4), July–August, 44–67.
4. Zhang X and Tao X, 'Smart textiles (1): passive smart', *Textile Asia*, 2001, June, 45–49.
5. Zhang X and Tao X, 'Smart textiles (2): active smart', *Textile Asia*, 2001, July, 49–52.
6. Zhang X and Tao X, 'Smart textiles (3): very smart', *Textile Asia*, 2001, August, 35–37.
7. Tao X (ed), *Smart Fibres, Fabrics and Clothing*, Woodhead Publishing Limited, Cambridge, England, 2001.
8. Mann S, 'Smart clothing: the shift to wearable computing', *Commun. Assoc. Computing Machinery*, 1996, **39**(8), 23–24.
9. Lukowicz P, Büren T V, Junker H, Stäger M and Tröster G, 'WearNET: a distributed multi sensor system for context aware wearables', In *Proceedings 4th International Conference on Ubiquitous Computing*, Göteborg, Sweden, 2002, Springer Verlag, 361–370.
10. Xybernaut Corporation, www.xybernaut.com
11. Gemperle F, Kasabach C, Stivoric J, Bauer M and Martin R, 'Design for wearability', In *Proceedings 2nd International Symposium on Wearable Computing*, Pittsburgh, USA, 1998, 116–122.
12. Lukowicz P, Anliker U and Tröster G, 'The WearARM modular, low-power computing core', *IEEE Micro*, 2001, May–June, 16–28.
13. Rantanen J, Impiö J, Karinsalo T, Malmivaara M and Reho A, 'Smart clothing prototype for the arctic environment', *Personal and Ubiquitous Computing*, 2002, **6**, 3–16.
14. '*ICD+ and Wearable Electronics*', Published: http://www.design.philips.com/pressroom/pressinfo/article/icd.html, 2000.
15. Coyle M, 'The Lifeshirt sytem: bringing high-tech patient monitoring from hospital to home', In *Proceedings 2nd International Avantex Symposium*, Frankfurt, Germany, 2002.
16. Kuhn H H and Child A D, 'Electrically conducting textiles', In Skotheim T A, Elsenbaumer R L and Reynolds J R (eds), *Handbook of Conducting Polymers*, 1998, Chapter 35, 993–1013.
17. Meoli D and May-Plumlee T, 'Interactive electronic textile development: a review of technologies' *J. Text. Apparel, Technol. Management*, 2002, **2**(2), 1–12.
18. Colortronics Inc., www.colortronics.com
19. Post E R, Orth M, Russo P R and Gershenfeld N, 'E-Broidery: design and fabrication of textile-based computing', *IBM Systems J.*, 2000, **39**(3/4), 840–860.
20. Gorlick M M, 'Electric suspenders: a fabric power bus and data network for wearable digital devices', In *Proceedings 3rd International Symposium on Wearable Computers*, San Francisco, USA, 1999, 114–121.
21. Park S, Mackenzie K and Jayaraman S, 'The wearable motherboard: a framework for personalized mobile information processing (PMIP)', In *Proceedings ACM/IEEE Design Automation Conference*, New Orleans, USA, 2002, 170–174.

22. Marculescu D, Marculescu R and Khosla P, 'Challenges and opportunities in electronic textiles modeling and optimization', In *Proceedings ACM/IEEE Design Automation Conference*, New Orleans, USA, 2002.
23. Estrin D, Reinmann G, Srivastava M and Sarrafzadeh M, 'Reconfigurable fabric', In *Proceedings ACM/IEEE Design Automation Conference*, New Orleans, USA, 2002.
24. Kallmayer C and Griese H, 'Fabric-based communication', *mstnews, Internat. Newslett. Micro-nano Integration*, 2002, **2**, 13–15.
25. Hum A P J, 'Fabric area network – a new wireless communications infrastructure to enable ubiquitous networking and sensing on intelligent clothing', *Computer Networks*, 2001, **35**, 391–399.
26. Massey P J, 'Fabric antennas for mobile telephony integrated within clothing', In *Proceedings London Communications Symposium*, 2000, on-line.
27. Leftly S and Jones D, 'SOFTswitch technology: the future of textile electronics', In *Proceedings 2nd International Avantex Symposium*, Frankfurt, Germany, 2002.
28. Swallow S S and Thompson A P, 'Sensory fabric for ubiquitous interfaces', *Internat. J. Human-Computer Interaction*, 2001, **13**(2), 147–159.
29. Farringdon J, Moore A J, Tilbury N, Church J and Biemond P D, 'Wearable sensor badge and sensor jacket for context awareness', In *Proceedings 3rd International Symposium on Wearable Computers*', 1999, 107–113.
30. Rix R, 'High-tech textiles – future perspectives: diverse applications of Gorix electro-conductive textiles', In *Proceedings 2nd International Avantex Symposium*, Frankfurt, Germany, 2002.
31. Tao X, Tang L, Du W and Choy C, 'Internal strain measurement by fibre Bragg grating sensors in textile composites', *Composites Sci. Technol.*, 2000, **60**, 657–669.
32. El-Sherif M A, Yuan J and MayDiarmid A, 'Fibre optic sensors and smart fabrics', *J. Intelligent Mater. Systems and Struct.*, 2000, **11**, May, 407–414.
33. Bar-Cohen Y, 'Electroactive polymers as artificial muscles – reality and challenges', In *Proceedings 42nd AIAA Structures, Structural Dynamics and Materials Conference*, Seattle, USA, 2001.
34. Gregory R V, Hanks T and Samuels R J, 'Chameleon fibers: dynamic color change from tunable molecular and oligomeric devices', *National Textile Center Research Briefs*, 2001, June, 6–7.
35. De Rossi D, Della Santa A and Mazzoldi A, 'Dressware: wearable hardware' *Mater. Sci. Eng.*, 1999, **7**, 31–35.
36. Deflin E, Weill A and Koncar V, 'Communicating clothes: optical fiber fabric for a new flexible display', In *Proceedings 2nd International Avantex Symposium*, Frankfurt, Germany, 2002.
37. Lu W, Fadeev A G, Qi B, Smela E, Mattes B R et al., 'Use of ionic liquids for pi-conjugated polymer electrochemical devices', *Science*, 2002, August, **297**, 983–987.
38. Hahn R and Reichl H, 'Batteries and power supplies for wearable and ubiquitous computing', In *Proceedings 3rd International Symposium on Wearable Computers*, San Francisco, USA, 1999, 168–169.
39. Starner T, 'Human-powered wearable computing', *IBM Systems J.*, 1996, **35**(3/4), 618–629.
40. Lauterbach C, Strasser M, Jung S and Weber W, ' "Smart clothes" self-powered by body heat', In *Proceedings 2nd International Avantex Symposium*, Frankfurt, Germany, 2002.
41. Baps B, Eber-Koyuncu M and Koyuncu M, 'Ceramic based solar cells in fiber form', *Key Eng. Mater.*, 2002, **206**, 937–940.

10
Data transfer for smart clothing: requirements and potential technologies

JAANA RANTANEN and MARKO HÄNNIKÄINEN
Tampere University of Technology, Finland

10.1 Introduction

Miniaturisation of electronic components has made it possible to build small portable and handheld computer devices that can be carried almost anywhere and at any time. As a result, smaller and lighter devices having high processing capacity are available on the market. This equipment is becoming more wearable since components can be easily hidden inside clothing or embedded in a handbag, for example, and carried for long periods.

A wearable computer is a miniaturised version of a desktop computer that is carried during use. Consequently, a wearable computer is a mobile, fully functional, self-powered and self-contained computer.[1,2] The basic difference from desktop computers is the type of a user interface (UI), since mobility sets new requirements for usability. Wearable computers are intended for general data processing tasks, similar to their desktop counterparts. Basically, the use of the computer is moved to the actual surroundings of everyday life.

Another approach in wearable electronics is smart clothing. Smart clothes emphasise the importance of clothing while designing and implementing the wearable systems. Smart clothing applications are constructed using functional modules or intelligent fabric materials that are placed on or inside ordinary clothes.[3] The functional modules can be non-electrical, e.g. an integrated first aid kit, but in our view they are considered to be electronics. Electronic functional modules for smart clothing applications are positioning, communication and sensor systems, and different types of UI components, for example.

When constructing smart clothes, several functional modules are distributed to optimal locations on clothing according to application design and user comfort. Therefore, the weight and size of the system are adapted. The system distribution results in data transfer requirements between the different modules. Since clothing has to maintain its profound properties, such as washability and wearing comfort, we have to consider carefully suitable solutions for data transfer.

Data transfer for smart clothing 199

For communication between the different components of smart clothing applications, both wired and wireless technologies are applicable. Wired connections are practical in many cases, but they can cause inflexibility and add to the weight of the system. Wireless connections increase flexibility but also the complexity of a system. Currently, data transfer issues are a true challenge in wearable systems. An applied solution is often a compromise based on application requirements, operational environment, available and known technologies, and costs.

This chapter evaluates the variety of technologies used to realise data transfer in smart clothing applications. Most potential technologies are considered for further analysis. Also, several smart clothing prototypes are introduced, concentrating on their data transfer solutions.

10.2 Smart clothing concept model

In introducing the architecture and functionality of smart clothing and its relation to the environment, a concept model has been used. An individual human user is the centre point of the model that is illustrated in Fig. 10.1. The concept model combines different clothing layers with additional components needed to integrate intelligence into clothing. The main layers concerned with smart clothing are the skin layer and two clothing layers.

Physically the closest clothing layer for a human user is an underwear layer, which transports perspiration away from the skin area. The function of this layer is to keep the interface between a user and the clothes comfortable and thus improve the overall wearing comfort. The second closest layer is an intermediate clothing layer, which consists of the clothes that are between the underclothes and outdoor

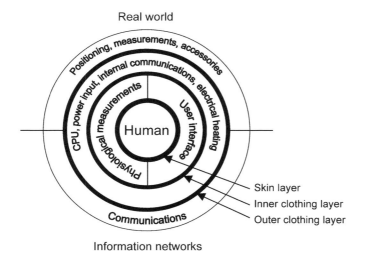

10.1 Concept model for smart clothing.

clothing. The main purpose of this layer is considered to be an insulation layer for warming up the body. The outermost layer is an outerwear layer, which protects a human against hard weather conditions.

Additional equipment that is needed to construct smart clothing systems can also be divided into layers in a similar manner. In our division, the underwear layer with additional components corresponds to the skin layer of smart clothing systems. In the same way, the intermediate clothing layer is associated with an inner clothing layer and the outermost layer with an outer clothing layer.

10.2.1 Smart clothing layers

The skin layer is located in close proximity to the skin. In this layer we place components that need direct contact with skin or need to be very close to the skin. Therefore, the layer consists of different UI devices and physiological measurement sensors. The number of the additional components in underwear is limited owing to the light structure of the clothing.

An inner clothing layer contains intermediate clothing equipped with electronic devices that do not need direct contact with skin and, on the other hand, do not need to be close to the surrounding environment. These components may also be larger in size and heavier in weight compared to components associated with underclothes. It is often beneficial to fasten components to the inner clothing layer, as they can be easily hidden. Surrounding clothes also protect electronic modules against cold, dirt and hard knocks.

Generally, the majority of electronic components can be placed on the inner clothing layer. These components include various sensors, a central processing unit (CPU) and communication equipment. Analogous to ordinary clothing, additional heating to warming up a person in cold weather conditions is also associated with this layer. Thus, the inner layer is the most suitable for batteries and power regulating equipment, which are also sources of heat.

The outer clothing layer contains sensors needed for environment measurements, positioning equipment that may need information from the surrounding environment and numerous other accessories. In Fig. 10.1, there are two different worlds (environments) presented that are in contact with the smart clothing. The term real world depicts the physical surroundings of smart clothing components that measure the environment. The term information networks represents the virtual environment accessed by communication technologies. The information networks can consist of communication with the external information systems, such as other network users and database servers.

10.2.2 Smart clothing implementation model

Generally, smart clothes are intended for very specific applications. Therefore, the intelligence is usually implemented using only a few selected components. For

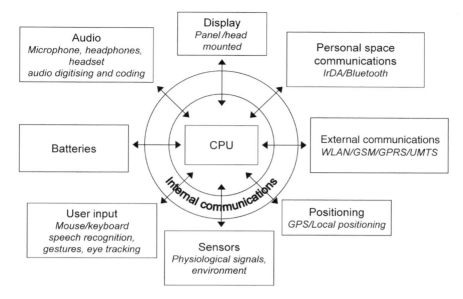

10.2 Block diagram of smart clothing components.

reference, a generic implementation architecture for smart clothing systems is depicted in Fig. 10.2, showing a number of different types of component. The necessary components are CPUs, various UI devices, power management equipment and data transfer components. The rest of the components vary according to the application requirements of the smart clothing system.

Central processing unit

The heart and brain of the smart clothing system is the CPU, where capacity varies according to the computing task. Often in smart clothing applications, small 8 to 16-bit microcontrollers are used.[3,4] In comparison, wearable computing applications usually utilise more powerful processors with speeds of up to 1 GHz.[5-7] The CPU module itself may be a combination of several microcontroller units that are distributed at several locations in the clothing.

User interfaces

UIs in smart clothing consist of several types of input and output devices for information feeding and selection. It is clear that devices suitable for desktop computing cannot be used with smart clothing applications. The ordinary keyboard and mouse have to be replaced by more suitable devices and new input/output concepts must be created. An example of a new innovative input device is the so-called Yo-Yo, which combines a display with a feeding and selecting system.[3]

Alternative input methods are pen-based inputs, gestures, eye movements and speech recognition inputs. The last is a very promising method since it allows hands free operation. Output devices consist of components that give feedback from the function of the clothing or from external actions. These include, for example, displays, loudspeakers, lights and haptic feedback devices. Commonly displays are small liquid crystal displays embedded in a suitable place in the clothing. Obtrusive head mounted displays are suitable for special applications such as protective clothing incorporating a helmet.

Power management

The most important design rule for power management in smart clothing is to minimise the power consumption. Batteries are heavy and thus difficult to place inside smart clothes. A centralised power source is easier for recharging purposes, but leads to wiring requirements for power transfer. A currently available solution is to use Li–polymer batteries that are thin and have a good power capacity. Alternative methods are also used such as kinetic energy and piezoelectric materials.[8,9]

10.3 Data transfer in smart clothing

A number of different wired and wireless data transfer technologies are applicable for the requirements placed by smart clothing. The communication model for data transfer in smart clothing and the potential technologies are discussed in the following sections.

10.3.1 Communication layers

Communications in smart clothing are divided into three different data transfer types. The communication model for smart clothing applications is illustrated in Fig. 10.3. First, the internal communication refers to the data transfer between the separate components of a distributed smart clothing implementation. This includes, for example, the data collected from physiological sensors and input/output messages through the UI. As the name implies this communication occurs inside clothes and between different smart clothing layers.

Second, external communication is needed for the data transfer between smart clothes and the external information networks. In a general communication model, there is only one access point at a time enabling the communication. For example, this access point can be a network interface for a cellular data network. The external communication is more easily manageable owing to this single access point.

The third type of communication is called personal space communication. Personal space data exchange takes place in situations when internal communication

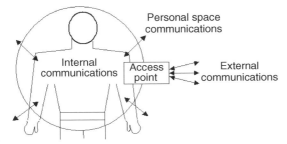

10.3 Relations between internal, external and personal space communications.

components initiate data transfer with an environment without a centralised access point, i.e. in an *ad hoc* manner. Personal space is the close proximity of the user, surrounding the human user while stationary or in motion. An example of such technology is a low range wireless link that can be utilised for both internal and external data transfer. The management of *ad hoc* external communications, consisting of possible several parallel dynamic connections, is a challenge for the system design.

10.3.2 Data transfer requirements

The communication requirements for smart clothing are firmly application dependent. A summary of different potential smart clothing applications and services with estimated data transfer requirements is presented in Table 10.1. The estimates presented for possible applications are related to the experience of wearable computer applications.

The transfer requirements can be divided into internal and external. In addition and within the personal space coverage, the external data transfer can be implemented using internal technologies. The internal transfer services are divided into local health and security related measurements, different services provided through a display and audio input/output UIs, and control type of input interfaces. Many of the services require or result in external communications between the smart clothing and its environment. For example, external transfer requirements are placed by the reception of a video or audio stream and text messages.

10.3.3 Wired solutions for internal data transfer

Wired data transfer is in many cases a practical and straightforward solution. Thin wires routed through fabric are an inexpensive and high capacity medium for information and power transfer. However, the detaching and reconnecting of wires decrease user comfort and the usability of clothes. An advanced wired solution is to use conductive fibres to replace ordinary plastic shielded wires. This makes smart clothing more like ordinary clothing. Also lightweight optical fibres are used

Table 10.1 Communication requirements estimates for smart clothing applications

Services/applications	Data components	Transfer requirements	Technologies
Internal			
Health and security	Physiological measurements	1–20 bit/s 10 s delay bound	Inductive coupling between separate clothing layers, conductive fibres within one clothing layer
UI voice (telephone)	Two-way audio stream	16–64 kbit/s 0.2 s delay bound	Wired and wireless headsets
UI video conferencing	One/two-way compressed low bit rate video Two-way audio stream	128–512 kbit/s 0.2 s delay bound	Cable or wireless link
UI audio streaming	One-way high quality audio	128–256 kbit/s 0.5 s delay bound	Cable or wireless link
UI video streaming	One-way high quality video	512 kbit/s–20 Mbit/s 1 s delay bound	Cable or high-speed wireless link
UI control (input)	One-way control messages (two-way with feedback devices)	0.1 kbit/s 0.1s delay bound	Cable or wireless link
External			
Web browsing	Web objects: 0.1–10 kB	256 kbit/s 0.5 s delay bound	Wireless high-speed networks
Email	1–5 kB (text only)	9.6 kbit/s 1 min delay bound	Wireless cellular telephone data Wireless high-speed networks
File transfer	1–10 MB files	1–10 Mbit/s 1 min delay bound	Infrared, wired high-speed networks, wireless high-speed networks
Real time media streaming	Two-way audio/video stream	62–20 Mbit/s 0.2–2 s delay bound	Real-time video and audio streaming
Network games	Two-way control messages 500 B	100 kbit/s 0.50 ms	Wireless high speed
Chat	100–200 character messages	1 kbit/s 2 s delay bound	Cellular telephone short message service
Positioning	Continuous measuring of radio signals	N/A	GPS receiver, local area radio positioning

in wearable applications, but their function has been closer to a sensor than a communication medium.[10,11] Optical fibres are commonly used for health monitoring applications and also for lighting purposes, e.g. in shoes.[12]

Cables

Wired communication implemented by plastic shielded cables is an inexpensive, high capacity and reliable data transfer method. Thin cables can be integrated or embedded inside clothing without affecting its appearance. However, wires form inflexible parts of clothing, thus decreasing the wearing comfort.[3] The cold winter environment especially stiffens the plastic shielding of cables. In hard usage and in cold weather conditions, cracking of wires also becomes a problem.[3]

The connections between the electrical components placed on different pieces of clothing are another challenge when using wires. During dressing and undressing, connectors should be attached or detached, decreasing the usability of clothing. Connectors should be easily fastened (or automatically fastened without special user attention), resulting in the need for new connector technologies.

Electrically conductive fibres

A potential alternative to plastic cables is to replace them with electrically conductive fibres. Conductive yarns twisted from fibres form a soft cable that naturally integrates in the clothing's structure keeping the system as clothing-like as possible. Fibre yarns provide durable, flexible and washable solutions. Electrically conductive yarns are either pure metal yarns or composites of metals with other materials. In composites other materials may, for example, provide strength or weight savings compared to pure metals. Metal clad aramid fibres are an example of strength solutions, which provide good electrical conductivity owing to a copper, silver or nickel coating.[13] A sophisticated solution would be to knit electrically conductive fibre yarns directly into cloths to form natural communication channels. In this way it is possible to construct wearable platforms, which already contain internal communication; only application-specific electrical components need to be added. However, conductive yarns are often used in the same way as plastic shielded cables.

Although this sounds easy, there are a few problems that slow down the usage of conductive fibres in clothing. The first problem is due to the lack of natural insulation material in conductive fibre yarns. Unshielded yarns can also conduct from their surface and this can cause unwanted short circuits when separate yarns are in touch with each other. Also conductive fibre yarns in close proximity, exposed to sweating or to other conductive material between the yarns may cause unwanted electrical conduction.

A possible solution is to embed conductive yarns inside waterproof tape. Tape shielding protects fibres against interferences from the outside world and acts as an

Table 10.2 Fibre yarn connection methods

Test fibre yarn	Connection type	Connection material
Bare Aracon®	Surface mount	Tin–lead solder
Shielded Aracon®	Surface mount	Tin–lead solder
Shielded Aracon®	Leading-through	Tin–lead solder
Shielded Aracon®	Leading-through	Conductive adhesive
Shielded Aracon®	Leading-through	Conductive adhesive, silicon elastomer
Bekinox	Leading-through	Tin–lead solder
Bekinox	Leading-through	Conductive adhesive, silicon elastomer

insulator. An example of this kind of shielding is Gore-Seam®, which is used to patch the holes made by the sewing machine and to ensure impermeability.[14] However, this adds working phases during production. Some conductive fibre yarns are also protected by a plastic shell. This kind of protection makes the yarn more like ordinary cable and lessens the clothing-like properties.

The second problem is due to the reliable connections of conductive fibre yarns. Ordinary cables can be soldered directly to printed circuit boards, but the structure of the fibre yarn is more sensitive to breakage near the solder connections. Protection materials that prevent the movement of the fibre yarn at the interface of the hard solder and the soft yarn must be used. For this purpose, as an example, silicone elastomer intended for electronics can be applied.

Some preliminary tests for connections on conductive fibre yarns and printed wiring boards were made in a laboratory climate chamber in varying humidity and temperature conditions. The materials used were bare metal clad aramid fibres, shielded metal clad aramid fibres, and conductive fibres made from stainless steel.[13,15] The fibre yarns were fastened to test printed circuit boards using tin–lead solder or conductive adhesive.

A summary of the tested connection methods is illustrated in Table 10.2. The test profile was in accordance with the MIL-STD-202F standard. MIL-STD-202 standard establishes uniform methods for testing electronic and electrical component parts, including basic environmental tests to determine resistance to deleterious effects of natural elements and conditions surrounding military operations, and physical and electrical tests (http://www.dscc.dla.mil/Programs/MilSpec/ListDocs.asp?BasicDoc=MIL-STD-202 provides military standards for download). A failure in connection causes the voltage over the connection to become temperature dependent. Therefore, during the test, voltages over the test connections were measured. After these tests, all the connections were functional. In a further analysis we took both stainless steel yarns and the best soldered and adhesive joints made from aramid fibres. In Fig. 10.4 we can see that aramid fibres managed the test better than stainless steel fibres. The graphs of shielded aramid fibre yarns with adhesive joints, silicon elastomer and soldered leading-through joints follow

Data transfer for smart clothing 207

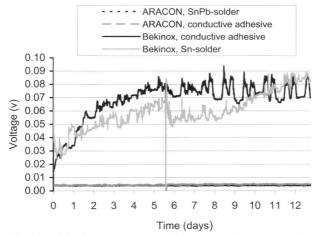

10.4 Humidity/temperature cycling test performance of stainless steel fibres and metal clad aramid fibres.

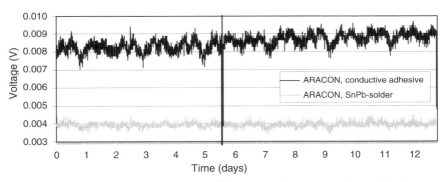

10.5 Humidity/temperature cycling test performance of metal clad aramid fibres with adhesive and solder joints.

almost the same path and the fluctuations of their contact voltages are smaller than the fluctuations of stainless steel yarns. These aramid fibre yarn joints are more accurately illustrated in Fig. 10.5. Connections made by Aracon® and tin–lead solder proved to be better than other connections. Joints made by conductive adhesive were worse than solder joints. Generally, shielded fibres endured much better mechanical stress than bare fibres.

10.3.4 Wireless technologies for data transfer

For wireless communications, dedicated external technologies for a wide range and internal technologies within personal space can both be utilised. These are discussed in the following section and summarised in Table 10.3.

Table 10.3 Wireless technologies for internal and external data transfer

Communications technology	Capacity	Service capability	Communication for smart clothing
GSM	43–171 kbit/s	Data and low quality voice	External
SMS/GSM	160 characters of text per message	Text, control messages	External
UMTS	144 kbit/–2 Mbit/s	High speed data, voice	External
IEEE and ETSI WLANs	11–54 Mbit/s	Data, QoS integration for real-time services is emerging	External
Bluetooth	1 Mbit/s	Data and low quality voice	Personal space
IEEE 802.15.1	Same as Bluetooth		
IEEE 802.15.3	55–100 Mbit/s	Emerging for multi-media services	Personal space
IEEE 802.15.4	20–250 kbit/s	Low rate data, control and diagnostics services, user interfaces	Internal
Low power RF	1–100 kbit/s	Control and sensors systems	Internal
Infrared	4 Mbit/s	High speed file transfers	Personal space

External data transfer

Digital cellular data networks, such as global system for mobile telecommunications (GSM), are a current technology for wide area voice and data services for smart clothing applications. GSM represents the latest technological state of current second-generation mobile networks. The system has been developed mainly for voice services, but it also possesses capabilities for general data transfer. Future extensions, such as universal mobile telecommunications systems (UMTS), should bring more bandwidth and enable new, more demanding applications in mobile wide area data networks.[16,17]

A basic service integrated into GSM is short message service (SMS). SMS enables text messages up to 160 characters to be sent between GSM terminals, and several messages can be concatenated. SMS has been found to be a suitable service for the data transfer requirements of varying control applications.

The third generation partnership project (3GPP) is continuing the effort started in the European Telecommunications Standards Institute (ETSI) to develop GSM with higher bit rates and new services. The first changes included an enhanced data rate of 14.4 kbit/s compared to the basic 9.6 kbit/s, still using a single voice call channel. High speed circuit switched data (HSCSD) and general packet radio system (GPRS) belong to the GSM 2.5 generation. HSCSD is an enhancement to the current circuit switched GSM data enabling the reservation of multiple voice channels for a single data transfer connection. With the maximum of four time slots, the symmetric data transfer rate is 57.6 kbit/s. Thus, data rates increase to the level of fixed telephone networks like the B-channel of ISDN (integrated services digital network). The first user terminals available on the market reach 43.2 kbit/s asymmetric data rates.

GPRS brings the GSM system closer to legacy data networks by providing packet access for a GSM terminal and packet switching based routing in the GSM infrastructure. The achieved data transfer rate depends on the availability of GSM channels in use per cell, giving a maximum rate of 171.2 kbit/s. The uplink and downlink data rates are asymmetric. GPRS allows a user to maintain a continuous virtual connection to the network, which facilitates several types of variable data rate services, including internet protocol (IP) based applications.

The work on a new modulation scheme and protocols for the GSM radio interface, called enhanced data rates for GSM evolution (EDGE), should extend the GSM data rates up to 348 kbit/s for HSCSD and GPRS data services, extending the GSM bit rates close to that of the 3G systems.

The terrestrial 3G UMTS will provide at least 144 kbit/s for full mobility applications in all environments, up to 348 kbit/s for macro cell and micro cell environments, and up to 2 Mbit/s short range coverage in micro and pico cell environments with limited mobility.

Wireless local area networks

Wireless local area networks (WLAN) can be utilised for high capacity communications within limited geographical areas, such as homes, offices and public hot-spot areas. Smart clothing applications are generally not expected to demand an office type of data communications, while WLAN is projected for a supplementary technology delivering third generation telecommunication services. Where WLAN infrastructure is available, WLAN provides reliable and low-cost data transfer, meeting most of the projected communication requirements for the external communications of smart clothing applications. Also, the miniaturisation of WLAN technology has proceeded, as WLANs have already been integrated into palm top computers.

The Institute of Electrical and Electronics Engineers (IEEE) standard 802.11 is currently the most widely used WLAN technology.[18,19] The system supports both direct *ad hoc* networking between users and infrastructure-based topology where WLAN access points manage the data transfer between terminals and provide a connection to fixed networks. One of the original 802.11 physical layers uses infrared technology while two of them are spread spectrum radios of the 2.4 GHz industrial, scientific, medical (ISM) band. All original physical layers provide up to 2 Mbit/s link rate.

The further development of IEEE 802.11 has proceeded with new physical layer technologies. The target has been to adapt the WLAN data rates to meet the wired LAN capacity better. Similarly, higher performance *ad hoc* networking has been approached. The 802.11b physical layer standard updates the link rate to 11 Mbit/s. The 802.11b is currently the most utilised physical layer of the existing 802.11 technologies.[20] The IEEE 802.11a standard specifies a physical layer for the 5 GHz band. The maximum link rate achieved with the 802.11a standard is 54 Mbit/s. Furthermore, the emerging 802.11g standard is specifying a 2.4 GHz band radio that has an equal capacity of 54 Mbit/s.

The support of WLAN data transfer quality of service (QoS) is being approached in terms of throughput with the development of higher rate radio layers. The QoS support for IEEE 802.11 WLANs is also being developed in the task group for the 802.11e standard. The purpose is to define procedures to support LAN applications with specific QoS requirements. Transport services for audio, voice and video applications have been appointed in the design requirements.

The European Telecommunications Standards Institute (ETSI) is developing WLAN technology specifications in the broadband radio access networks (BRAN) project. There are also specifications for interfacing existing wired networks.[21] ETSI high performance radio local area network (HIPERLAN) type-1 (HIPERLAN/1) specifies a 5 GHz band radio, with the maximum signalling rate of 23.5 Mbit/s for data transmission. HIPERLAN/1 has a fully distributed network topology. In addition, to extend the network coverage, HIPERLAN/1 utilises multi-hop relaying, in which intermediate terminals can forward received frames towards their final destination. Interconnection with a peer LAN is enabled with a bridging terminal.[22]

HIPERLAN type 2 (HIPERLAN/2) is a mobile short-range access technology for broadband networks, such as IP (over wired LANs). HIPERLAN/2 has a centralised network topology and mobile terminals communicate through an access point in a connection-oriented manner. A capability for *ad hoc* type direct communication between terminals is provided, but a central controller entity is still required to control the data transfer.

The HIPERLAN/2 physical layer technology is similar to the IEEE 802.11a, and the achieved link rate is similarly also 54 Mbit/s. Convergence layers adapt the core network technologies to the HIPERLAN/2. For each of the supported core networks, such as IEEE 802.3 (Ethernet), UMTS and IEEE 1394, a separate convergence layer is specified.

10.3.5 Internal and personal space data transfer

Potential operational frequencies for internal and personal space wireless data transfer are radio bands that do not require a specific licence, special permission or carry licence fees, such as the 2.4 GHz ISM band. In Europe, ISM bands are part of the frequencies allocated for short range devices (SRD). Table 10.4 summarises

Table 10.4 Potential WLAN frequency bands in Europe

Frequency band	Frequencies	Maximum power (EIRP)	Existing/expected technologies and applications
433 MHz ISM	433.05–434.79 MHz	1 mW	RFID technologies, baby monitors, cordless headphones, walkie-talkie phones
868 MHz SRD	868–870 MHz (with several sub band divisions)	5–500 mW	Cordless audio devices, radio microphones, general purpose telemetry, general purpose alarms
2.4 GHz ISM	2.4000–2.4835 GHz	100 mW	WLANs, WPANs
5 GHz HIPERLAN	5.150–5.350 GHz	200 mW	WLAN (indoor only)
	5.470–5.725 GHz	1W	WLAN
5 GHz ISM	5.725–5.875 MHz	25 mW	WLAN, WPAN
17 GHz HIPERLAN	17.1–17.3 GHz	100 mW	WPAN
60 GHz and higher ISM bands	61–61.5 GHz 122–123 GHz 244–246 GHz	N/A	Future development

EIRP = effective isotropic radiated power.

the available ISM bands in Europe, with different existing and emerging technologies and applications.

Potential wireless solutions for internal communications are inductive coupling, infrared and radio frequency (RF) technologies. Low-power RF is more flexible, but these technologies are generally proprietary, while interoperable solutions are emerging for the 2.4 GHz industrial ISM band. However, the standardisation for simpler wireless technologies, targeting at sensor applications is advancing.

Generally, wireless personal area network (WPAN) technology can form a multipurpose link for extending and delivering services to smart clothing applications. WPAN differs from WLAN mainly by non-functional requirements, such as cost and power consumption that favour smart clothing types of applications. Also, WPAN has a smaller operational area, lower data rate and fewer terminals per network compared to WLAN.

Bluetooth is the first available WPAN technology. The technology is a potential standard for low-range RF links, supporting both data and voice services. Bluetooth is a WPAN technology specified by an industry driven organisation called the Bluetooth Special Interest Group (SIG).[23] The main target of the technology is to replace a common serial cable with a wireless link. The non-functional requirements have been emphasised. Thus, the technology targets low cost, low power consumption and small size. Consequently, Bluetooth is expected to be integrated into numerous personal devices, such as palm top computers and mobile phones.

Home and industrial automation has also been a potential application type for such technology.

The Bluetooth radio operates on the same 2.4 GHZ ISM band as current WLANs. The achieved link rate is 1 Mbit/s. The achieved data rate is dependent on the utilised link and packet types. Two link types categorise Bluetooth services. The Asynchronous Connectionless (ACL) data link provides up to 721 kbit/s asymmetric data rate. The Synchronous Connection Oriented (SCO) voice link has a 64 kbit/s rate, and up to three voice connections can exist at the same time.[24]

A Bluetooth link has a centralised access control, but the network (piconet) is constructed automatically in an *ad hoc* fashion between a master node and up to seven slave nodes. In addition, a number of other slaves can be associated with the same piconet while being in a power save mode. Several piconets can form a scatternet, as a Bluetooth node can participate in several piconets at the same time.

The IEEE 802.15 working group standardises WPANs and short distance wireless data communications in general.[25] There are four task groups. Task group 2 is developing practices and mechanisms to facilitate the coexistence of 802.11 WLAN and 802.15 WPAN that operate on the same band. The other groups are developing WPAN technologies.

The first task group of 802.15 (802.15.1) has adopted a WPAN standard from the Bluetooth specification. The 802.15 task group 3 is developing a new standard for a high rate WPAN technology. The target rate has been over 20 Mbit/s, which enables a wider range of personal area applications, such as image and multimedia transfer for consumer electronics appliances. The draft standard defines a physical layer with up to 55 Mbit/s link speed for the 2.4 GHz radio band and a link protocol destined to support multimedia applications. Furthermore, study group 802.15.3a has been recently established to specify an alternative higher rate physical layer targeting an over 100 Mbit/s link rate.

The IEEE 802.15 working group 4 is developing specifications for low data rate WPANs. Target applications are low complexity embedded systems that require a long battery life, such as sensors, interactive toys, remote controls and home automation devices. The draft 802.15.4 standard proposes a single physical layer, but the layer can operate on two different frequency bands. Furthermore, the lower band is either the 868 MHz SRD band in Europe or the 902 MHz ISM band in USA. The higher band is the 2.4 GHz ISM. The targeted link rate is 20 kbit/s for the lower two bands and 250 kbit/s for the 2.4 GHz band.

Infrared communication is widely integrated, for example, in remote controllers, mobile phones, notebooks and digital cameras. Infrared is a low-cost solution but the line-of-sight requirement is problematic especially in devices carried on the body. However, for personal space communication a short-range infrared data transfer can turn out to be a practical and inexpensive solution. Infrared data association (IrDA) has standardised infrared communication with IrDA DATA and IrDA control standards.[26] IrDA DATA defines data transfer for a universal data port, while IrDA control defines data transfer for simple controlling devices such as keyboards and mice. IrDA communication is intended to replace point-to-point cable and it supports data rates up to 4 Mbit/s.

An alternative to infrared communication is low-range and low-power RF communication, which provides more freedom and flexibility to users, since no line-of-sight is needed. The 433.92 MHz and 868 MHz ISM bands are currently used by similar systems, for example in home and office automation applications. As Bluetooth has not emerged as quickly as expected, these low-power RF modules are capturing markets. On the other hand Bluetooth may also turn out to be an expensive and complex solution for the most simple sensor applications.

A recent standardisation approach in IEEE for low power RF systems has been taken in the P1451.5 working group for wireless sensor standards.[27] The purpose for the group is to develop an open standard for wireless transducer communication that can accommodate various existing wireless technologies. The emerging work of the IEEE P1451.5 working groups seems to be very promising for the smart clothing data transfer requirements.

Several RF components suppliers are providing modules with programmable transmission power and with several frequency alternatives. RF link implementations tested at Tampere University of Technology in the Institute of Electronics include ChipCon's circuits CC1000 and CC900, and Nordic VLSI's nRF401circuits.[28–30] The purpose of the study was to find suitable RF link solutions for smart clothing applications. The most important selection criteria were low power consumption and reliable data transfer. In the study, the range, transmission power and the power consumption of the RF circuits were measured. Based on these measurements power consumption per metre was calculated. In these tests CC1000 fulfilled the set requirements best. In addition CC1000 was easy to use and only a few additional components were needed for implementation.

Radio frequency identification technology (RFID) is used in various tracking and identification applications, including smart cards, access control, logistics, sport events, electronic article surveillance and animal identification. A large number of these RFID systems work using an inductive coupling principle in the data transmission.[31] This means the transfer of energy from one circuit to another by means of mutual inductance between these circuits. Another data transfer method is based on propagating electromagnetic waves. These systems provide higher data transfer rates and reading ranges than inductive coupled systems. The main components of the RFID systems are a reader and a tag. In passive systems the reader provides the necessary energy for the operation of the tag, whereas in active systems a separate power source for the tag is needed.

Different transmissions frequencies are classified into three classes; low frequency between 30 and 300 kHz, high frequency from 3 to 30 MHz and ultra high frequency from 300 MHz to 3 GHz and the microwave range above this.[31] These systems can be divided further according to the operation range. In close coupling systems the reader and the tag must be integrated or placed one upon the other and the operation range is up to 1 cm. Remote coupling and long-range systems can operate in the range up to 1 and 10 m, respectively. Generally, higher frequency systems can operate in longer reading ranges.

RFID systems provide low-cost, non line-of-sight wireless links, which makes this technology an attractive alternative to smart clothing applications. For internal communication, low frequency and close coupling RFID systems can be a potential solution for data transfer between different pieces of clothing.

For RFID, communication is generally restricted to the surfaces of clothes, since the strength of the magnetic field decays according to the increasing perpendicular distance from the centre of the reader's loop antenna.[32] This is suitable for smart clothing applications, but in general it is a factor that limits the reading range of RFID systems. Data transfer occurs in small overlapping areas of the reader and the tag. In these systems, inductive coupling works as a connector between different layers of clothing. Remote coupling and long-range systems can be used in certain control types of applications for personal space and external communication purposes.

Usually inductive coupled systems work as single coil systems. However, an example of a new network solution is being developed at Starlab.[33] This fabric area network (FAN) is the first reported inductive coupled communication medium for smart clothing applications. It uses 125 kHz RFID technology and data transfer is limited to a distance smaller than 2 cm. Nodes are routed from a central controlling base station to several locations on the clothing. These nodes can then communicate with tags that come near them.

In addition to inductive coupling, capacitive coupling can also be used in close coupling applications. Capacitive coupling takes place between two coupling surfaces, which are located within a short distance of each other. This coupling method is used in close coupling smart card applications.[31] An interesting approach to capacitive coupling has also been introduced by Zimmerman.[34] In this implementation of PAN, small currents are induced through the human body. The system contains a transmitter and a receiver with a pair of electrodes. The transmitter capacitively couples a small current through the body to the receiver. The return path is the earth (ground) and therefore the body needs to be electrically isolated. Watches, name badges and pocket inserts are examples of devices that can use capacitive PAN technology for data transfer.

10.4 Implementations for communication

In this section, implementations of a wearable computer vest and three smart clothing prototypes are introduced. The emphasis is on the data transfer solutions.

10.4.1 Wearable computer vest

A wearable computer vest, constructed from commercial components in 1998, is illustrated in Fig. 10.6. This wearable computer solution was our first prototype at

Data transfer for smart clothing   215

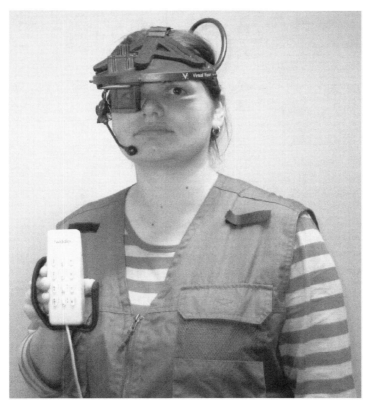

10.6 First wearable computer assembly at Tampere University of Technology.

Tampere University of Technology and later solutions have been partly based on the experiences acquired. The platform for the computer system is the vest, which contains several pockets for additional components. The function of the application is to be a general helper, providing a variety of data processing functions in everyday situations. The computer is as mobile as its user. The CPU in this solution is Via II, which is integrated into the back of the vest.[5]

A head mounted display V-Cap 1000 manufactured by Virtual Vision Inc. is connected to the CPU module with a thick and inflexible cable. The cable is a bit clumsy, although now there are also wireless solutions available on the markets. The input device in the configuration presented in Fig. 10.6 is a chording keyboard called Twiddler, which allows one hand operation.[35] In the vest, the connector cable for Twiddler is a problem and a wireless solution would be better. The application has also two different positioning systems. First, we use an infrared-based tag system indoors.[36] Second, we can use global positioning system (GPS) outdoors. Continuous connection to the internet and other information networks is

10.7 Reima smart clothing prototype for the arctic environment.

achieved using a WLAN (Lucent WaveLAN) network adapter within the premises of the Institute of Electronics. In other places we can use a GSM card phone (Nokia).[37]

10.4.2 Reima survival suit

A Reima smart clothing prototype illustrated in Fig. 10.7 is intended to assist survival in the arctic environment. The prototype is suitable for several activities in a harsh winter environment, but the special target user group is experienced

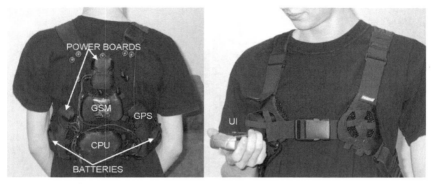

10.8 Support structure for Reima smart clothing prototype.

snowmobile users. The prototype suit consists of underclothes, a supporting vest and an actual snowmobile jacket and trousers.

The function of the prototype is to prevent accidents and, on the other hand, help users to survive longer in the case where an accident has already happened. The suit is capable of acquiring information of the wearer's health, location and movements. With several types of integrated sensors, it is possible to monitor the user's condition and position. If a user encounters an accident or another abnormal situation, the suit will inform an emergency office or use another preselected phone number via the SMS of the GSM modem. The user can also send an emergency message him- or herself. The message contains the current coordinates of the user's position, the data from the user and the environment measurements. The coordinates are acquired using GPS.

The three-layer structure of the suit allows data transfer between different pieces of clothing. Special underwear (i.e. the skin layer) incorporate a heart rate sensor and electrical heating panels in the end of the sleeves. In this prototype, a special supporting structure forming an inner clothing layer is embedded inside the snowmobile jacket. This supporting vest illustrated in Fig. 10.8 contains most of the electrical components.

The outerwear contains a few environment measurement sensors and UI devices. The UI consists of three separate devices: a small loudspeaker on the collar of the outerwear, a light feedback device in the sleeve of the jacket and the Yo-Yo interface in the front pocket of the jacket. The UI is physically placed on the outerwear since we have to operate with it continuously in the cold winter environment.

Cables were used for internal data transfer between the CPU and other components. Selection between different types of cables was done so that the plastic shielding would also endure in harsh environments. Since cables were used in data transfer between different pieces of clothing, some extra connectors were needed. However, this was the only possible solution, since the heavy batteries needed for heating were not practical in the structure of the underwear. Cables

10.9 Sensor shirt and heating jacket.

were also used for data transfer between the CPU and the Yo-Yo. In contrast, wireless data transfer is used between the heart rate sensor and the CPU. This is implemented by using a low frequency magnetic field.

Since the UI is in the breast pocket it is subject to weather effects. Therefore, a wireless link between the UI and the CPU would be more reliable. A number of wires are placed into the supporting vest, but its structure is a natural platform for fastening cables. In addition, it is not possible to replace these cables with fibres, and replacing them with wireless links could cause them to interfere with each other. In this case, wires were the best solution. Further information on the smart clothing prototype can be obtained from Rantanen *et al.*[3]

10.4.3 Heating jacket and sensor shirt

An electrically heated smart clothing prototype is illustrated in Fig. 10.9.[38] On the left is the sensor shirt, which is a platform for sensors and essential measurement electronics. The heating control is based on the physiological measurements made by the shirt. On the right is the actual heating jacket, which contains heating elements, voltage regulation electronics and power control electronics. The main target of the whole system is to help the user to reach a condition of thermal comfort. The sensor shirt is also functional without the heating jacket.

The temperature sensors measure skin surface temperature at ten different

10.10 A PPM vest.

places. In addition, skin conductivity and respiration can be measured. Data transfer between the measurement electronics and the temperature sensors is implemented by electrically conductive fibre yarns. The same yarn is also used for skin conductivity electrodes. The fibre yarn makes the shirt soft and comfortable to wear, but connections between the yarns and the temperature sensors are unreliable. The shielding rubber that covers the connections is too hard for yarns and causes breakage of the structure of yarns. The measurement electronics boards are located so close to each other that cables can be easily used for data transfer between them.

A test UI for the application is a personal digital assistant (PDA), which is connected to the CPU by a serial cable. This cable causes disturbance and should be replaced by a suitable wireless link. Communication inside the jacket is implemented by cables.

10.4.4 Personal position manager

The personal position manager (PPM) uses a fishing vest as a platform for electrical devices. PPM is an application that is particularly intended for outdoor activity, such as fishing and hiking.[39] The basic function of the system is to record good fishing places, so that fishermen can find places for the best catch again. It is also possible to request the route to a desired destination point that was saved earlier into system's memory, to record continuously the last 10 km of route travelled and to follow a route travelled. In this implementation, the requirements of easy usage and low power consumption guided the design process. A PPM is illustrated in Fig. 10.10.

The system consists of a GPS for position tracking, a CPU to control the function of the vest, a power source and two types of UI. PDA with implemented UI is used for navigation and a touch button UI is used for marking interesting places. The majority of the components are hidden between the outerwear cloth and lining of the fishing vest. Only the UIs are placed on the surface side of the vest so that they are accessible all the time. Special pockets for components are sewn inside the vest and therefore cables are the natural solution to connect the distributed components to each other. At the beginning, we also used a cable between CPU and PDA UI. However, it proved difficult to use since the connector easily detached and the extra cable hanging from the pocket was a negative alternative visually. We have now replaced the cable by a low-power RF link. The RF communication is implemented by using ChipCon's CC1000 circuit.[28]

10.5 Summary

A number of wired and wireless data transfer technologies are available for smart clothing applications. For wearability, conductive fibres are seen as the most suitable wired solution, while ordinary cables provide high reliability. Low-power wireless connections provide increased flexibility and also enable external data transfer within the personal space. Different existing and emerging WLAN and WPAN types of technologies are general purpose solutions for the external communications, providing both high speed transfer and low costs. For wider area communications and full mobility, cellular data networks are currently the only practical possibility. Experiences with prototypes have shown the operability and potential of smart clothing, and also indicated the need for research work on new technologies and usability.

10.6 References

1. Mann S, 'Smart clothing: the shift to wearable computing', *Commun. ACM*, 1996, **39**(8), 23–24.
2. Bass L, Kasabach C, Martin R, Siewiorek D, Smailagic A and Stivoric J, 'The design of a wearable computer', *Conference on Human Factors in Computing Systems (CHI)*, Atlanta, GA, USA, ACM, 22–27 March 1997, 139–146.
3. Rantanen J, Impiö J, Karinsalo T, Malmivaara M, Reho A, Tasanen M and Vanhala J, 'Smart clothing prototype for the arctic environment', *Personal and Ubiquitous Computing*, 2002, **6**(1), 3–16.
4. Kukkonen K, Vuorela T, Rantanen J, Ryynänen O, Siili A and Vanhala J, 'The design and implementation of electrically heated clothing', *5th International Symposium on Wearable Computers (ISWC)*, Zürich, Switzerland, 8–9 October 2001, 180–181.

Data transfer for smart clothing   221

5. Homepage of ViA Inc., http://www.via-pc.com/, 1 November 2002.
6. Homepage of Xybernaut Corporation, http://www.xybernaut.com/, 1 November 2002.
7. Homepage of Charmed Technology Inc., http://www.charmed.com/, 1 November 2002.
8. Homepage of Seiko Kinetic Watches, http://www.seiko-kinetic-watches.com/, 1 November 2002.
9. Kymissis J, Kendall C, Paradiso J and Gershenfeld N, 'Parasitic power harvesting in shoes', *2nd International Symposium on Wearable Computers (ISWC)*, Pittsburgh, PA, USA, 19–20 October 1998, 132–139.
10. Lind E J, Jayaraman S, Eisler R and McKee T, 'A sensate liner for personnel monitoring applications', *1st International Symposium on Wearable Computers (ISWC)*, Cambridge, MA, USA, 13–14 October, 1997, 98–105.
11. Lee K and Kwon D, 'Wearable master device using optical fiber curvature sensors for the disabled', *International Conference on Robotics & Automation*, Seoul, Korea, 21–26 May 2001, 892–896.
12. Homepage of Lumitex lightning applications, http://www.lumitex.com/other_fiber_optic_lighting.html, 1 November 2002.
13. DuPont Advanced Fibre Systems, *Product Bulletin on Aracon® Brand Metal Clad Fiber Type XS0200E-025*, Wilmington, DE, USA.
14. Homepage of Gore Military Fabrics, http://www.goremilitary.com, 16 October 2002.
15. Bekaert N V and Bekaert S A, *Fibre Technologies*, Brochure of Bekonox® and Bekitex® Metal Fibres for Conductive Textiles, Zwevegem, Belgium.
16. Hännikäinen M, Hämäläinen T, Niemi M and Saarinen J, 'Trends in personal wireless data communications', *Computer Communications*, 2002, **25**(1), 84–99.
17. Homepage of the 3rd Generation Partnership Project (3GPP), http://www.3gpp.org/, 26 October 2002.
18. Homepage of the IEEE 802.11 Working Group, http://grouper.ieee.org/groups/802/11, 23 March 2002.
19. ISO/IEC 8802-11, 'Wireless LAN Medium Access Control (MAC) and Physical Layer (PHY) specifications', 1999.
20. IEEE Std 802.11b-1999, (Supplement to ANSI/IEEE Std 802.11, 1999 Edition), *Higher-Speed Physical Layer Extension in the 2.4 GHz Band*, September 1999.
21. Homepage of the ETSI BRAN Project, http://www.etsi.org/bran/, 26 April 2002.
22. EN 300 652 V1.2.1, *Broadband Radio Access Networks (BRAN); High Performance Radio Local Area Network (HIPERLAN) Type 1; Functional specification*, 1998.
23. Homepage of the Bluetooth SIG, http://www.bluetooth.com/, 26 April 2002.
24. Bluetooth S I G, *Specification of the Bluetooth System, Core*, Version 1.1, 1082 pages, February 2001.
25. Homepage of the IEEE 802.15 Working Group, http://grouper.ieee.org/groups/802/15, 23 October 2002.
26. Homepage of the Infrared Data Association (IrDA), http://www.irda.org, 26 October 2002.
27. Homepage of the IEEE P1451.5 Working Group, http://grouper.ieee.org/groups/1451/5/, 26 October 2002.
28. ChipCon AS, Oslo, Norway, *Datasheets of CC1000 Single Chip Very Low Power RF Transceiver*.
29. ChipCon AS, Oslo, Norway, *Datasheets of CC900 Single Chip High Performance RF Transceiver*.

30. Nordic VLSI ASA, Tiller, Norway, *Data Sheets of 433 MHz Single Chip Transceiver.*
31. Finkenzeller K, *RFID Handbook – Radio Frequency Identification Fundamentals and Applications*, England, Wiley & Sons, 1999.
32. Chen S C Q and Thomas V, 'Optimization of inductive RFID technology', *IEE Symposium on Electronics and the Environment*, Denver, CO, USA, 7–9 May 2001, 82–87.
33. Hum A P J, 'Fabric area network – a new wireless communications infrastructure to enable ubiquitous networking and sensing on intelligent clothing', Special Issue on *Pervasive and Computer Networks*, 2000, **35**(4), 391–399.
34. Zimmerman T G, 'Personal area networks: near-field intrabody communication', *IBM Systems J.*, 1996, **35**(3/4), 609–617.
35. Homepage of Handykey Corporation, http://www.handykey.com/, 1 November 2002.
36. Häkkinen T and Vanhala J, 'Infrared tag equipment for identification and indoor positioning applications', *International Conference on Machine Automation (ICMA)*, Osaka, Japan, 27–29 September 2000, 283–288.
37. Homepage of Nokia Card Phones, http://www.forum.nokia.com/, 1 November 2002.
38. Rantanen J, Vuorela T, Kukkonen K, Ryynänen O, Siili A and Vanhala J, 'Improving human thermal comfort with smart clothing', *IEEE Conference on Systems, Man, and Cybernetics*, Tucson, AZ, USA, 7–10 October 2001, 795–800.
39. Rantanen J, Alho T, Kukkonen K, Vuorela T and Vanhala J, 'Wearable platform for outdoor positioning', *4th International Conference on Machine Automation (ICMA)*, Tampere, Finland, 11–13 September 2002, 347–357.

11
Interaction design in smart textiles clothing and applications

SHARON BAURLEY
University of the Arts London, UK

11.1 Introduction

This chapter suggests that shifts in the textiles, electronics, and information and communication technology sectors will give rise to the area of intelligent textiles and clothing. The technical textiles industry in the USA and the EU is growing. The western clothing market has segmented into two distinct areas: low-cost, high-volume and high-end specification goods, for example sports performance, and designer-led fashion. The textiles industries of the USA and the EU are focusing on technical textiles for high-specification products, partly in response to the approaching end of the 'Multi-Fibre Arrangement'. Since 1974, the world trade in textiles and garments has been governed by the Multi-Fibre Arrangement. This agreement provided the basis on which industrialised countries have been able to restrict imports from developing countries. It expired at the end of 2004. In recent years the market growth in clothing has been fuelled by the emergence of new fibres, new fabrics and innovative processing technologies. This trend, in which technical innovations in textiles will become more important than the fashion content itself, is set to continue. The market has also been boosted by changes in consumer lifestyles. Many of these new developments have come from the technical textiles industry. High tech fabrics must continue to cross the boundary into everyday fashion apparel as well as into home interior furnishings to meet the challenge of future lifestyle needs and consumer requirements. Products that win will be those that enhance life quality in some way, and have added value in terms of functionality and performance. As we advance into the knowledge age, objects and material technology are forecast to pervade our material environment. The market for technology products generally is growing. What consumers require of products is changing, gravitating towards higher-order needs that stimulate the intellect by providing experience and sensory and emotional fulfilment. Such products are set to become the new commercial imperatives of the developed world.

The integration of smart functionality into clothing and other textile products will fundamentally change the cultures of clothing and interior products. It will also radically alter the relationship that people have with them and, hence, the way these products are designed and the materials used to produce them are developed. This chapter highlights key developments in computing, such as ubiquitous computing and human–computer interaction. It suggests, through current research in fashion and textiles at Central Saint Martins College of Art and Design (CSM), London, that the design of textile products will converge towards computing and the field of human–computer interaction design. The areas in which textile materials might develop, the ways in which different sectors of industry will need to collaborate and possibly converge for these developments to occur and the implications for the development process are also highlighted.

11.2 Knowledge age: dematerialisation of information and communications technology and the rise of ubiquitous intelligence

We are advancing into a knowledge-based economy, where ideas and information mean capital, and access to information and communication systems are the drivers. Science is increasingly affecting all aspects of our lives through products and services, as the market for technology products expands. According to Philips Electronics, based in the Netherlands, 'Our environment of the future will consist of invisible interactive systems that will be embedded in our living spaces and clothing, creating an ambient intelligence that could form a natural part of our life' (Marzano and Arts, 2003). The silver and black plastic products that currently house electronics and computers are set to vanish as technology dissolves into our material environment, i.e. interiors, buildings, furniture and clothing. As technology becomes dematerialised and embedded within these hitherto dumb products, such products will become active and intelligent. They will be the future mediators of technology. Our contact with these everyday objects will become a central focus in our lives, facilitating new methods of accessing entertainment, knowledge and communication. The following paragraphs contain examples of new methods of accessing information that have been in development over the last twenty years.

The area of ubiquitous computing is involved with how people access the computer intelligence embedded within everyday objects and devices, whose user interface is intuitive. Mark Weiser, chief technologist at Xerox Parc research centre, Palo Alto, CA, USA, is the father of ubiquitous computing, which he has dubbed:

> the third wave in computing. First were mainframes, each shared by lots of people. Now we are in the personal computing era, person and machine staring uneasily at each other across the desktop. Next comes ubiquitous computing, or the age of calm technology, when technology recedes into the background of our lives. (Weiser, 1996)

IBM (International Business Machines) has identified:

> four major aspects of pervasive computing that appeal to the general population: Computing is spread throughout the environment; users are mobile; information appliances are becoming increasingly available; communication is made easier between individuals, between individuals and things, and between things. (Ark, 1999)

Pervasive computing is about systems that are embedded everywhere, which do not require the user to understand their inner workings.

There are various initiatives and research centres that address these shifts, for example the EU Disappearing Computer initiative, the IBM Almaden Research Centre, USA and the Massachussetts Institute of Technology, MIT Media Lab, USA. These initiatives examine the relationships we have with computer intelligence and how we interact with it. Also, there has been much activity in the last ten years in wearable computing driven by the computing sector. A good example of the ubiquitous computing vision is Smart Paper by Gyricon, which combines the best of modern computing technology with the best of established technology, the book. The vision of the paperless office has never materialised – in fact, we are using more paper now than ever before. This is because no one wants to read text off a computer display; the printed word on paper simply looks better. By coating paper with electronic ink, the contents of a book or a newspaper can change on command or continuously, when it is downloaded wirelessly.

Steven Mann is regarded as one of the inventors of wearable computing. A professor in the Computer Engineering Research Group at the University of Toronto, Canada, Mann has been working on his *WearComp* (wearable computer) invention for more than twenty years. He brought his inventions and ideas to MIT in 1991, which contributed to the foundation of the MIT Wearable Computing Project. Mann views wearables as a means of 'personal empowerment':

> Wearable computing facilitates a new form of human–computer interaction comprising a small body worn computer system that is always on and always ready and accessible. In this regard, the new computational framework differs from that of hand held devices, laptop computers and personal digital assistants. The *always ready* capability leads to a new form of synergy between human and computer, characterized by long-term adaptation through constancy of user-interface. (Mann, 1998)

Research in the area of wearable input devices that tries to address the notion of intuitive interaction, for example, the *Gesture Wrist* and *Gesture Pad* by Sony Computer Science Laboratories. 'These devices allow users to interact with wearable or nearby computers by using gesture-based commands' (Rekimoto, 2001). The researchers formulated a series of gestures that the wearable computer recognises as input signals. The gesture configurations, however, are not based on what might be considered intuitive human gestures, but on invented gestures.

The area of affective computing, pioneered by Roz Picard at MIT's Media Lab., has grown out of wearable computing. Affective computing is concerned with educating a computer (wearable or non-wearable) to recognise physical and physiological patterns and translate these into emotions. For example, 'expressions of emotion such as a joyful smile, an angry gesture, a strained voice or a change in autonomic nervous system activity such as accelerated heart rate or increasing skin conductivity' (Picard and Healy, 1997). By integrating sensors into clothing, a computer can sense a user's emotional status, enabling it to understand what its user wants and to effect responses that adjust to the user's patterns. Assumptions of emotional status are made on the basis of physiological behaviour, and much testing has been done on wearable computers (wearables) to test out emotion theories (an area of research in psychology, on which there are several classic theories) to attain a realistic assessment of the emotional meaning of physiological patterns. These tests are underpinned by the work of research psychologists who have been studying emotion and its relation to behaviour. Picard's team has developed a number of prototype affective wearables, such as earrings that measure blood volume pulses, sandals that gauge skin conductivity and glasses that check facial expressions.

From ubiquitous and wearable computing has arisen the field of human–computer interaction design. Interaction is the means by which users input changes to and receive feedback from an information technology (IT) system. Interaction design involves 'human cognition, context of use, platform of access, task analysis and user experience' (Macdonald, 2003). Interaction design comprises a distinct set of skills required to design the forms through which people can successfully use information technology in ways that are meaningful. 'As products and services are increasingly being created using information technology, interaction design is likely to become the key design skill of this century' (Macdonald, 2003). Examples of institutions across the globe engaged in human–computer interaction design include the Interaction Design Department Royal College of Art, UK, the Interactive Institute, Ivrea, Italy, the Things That Think Group, MIT, USA and the Almaden Research Centre, IBM, USA.

11.3 New commercial imperatives

Running in parallel with, and complementary to, the rise of ubiquitous computing and access to information are changing consumer requirements. The societies of the developed world are gravitating towards a culture that will be focused firmly on human senses. As we diverge from a purely material culture, a culture firmly focused on human senses is becoming epitomised by a requirement for more intensive experiences and higher order meanings. According to Maslow's hierarchy of needs theory, there are two levels of need, basic and meta, along which people constantly strive to move. Now that basic needs have been met in the developed world, people are striving to satisfy meta or higher-order needs. Meta needs include cognitive, aesthetic, self-actualisation and self-transcendent. Hence, the transition from making and marketing a product to developing non-tangible concepts that satisfy the demands of higher-order needs is underway and gaining momentum. Such needs include ideas, sensory and emotional fulfilment, cultural experiences and entertainment that stimulate the intellect. 'A new segment is emerging in the consumer market place, . . . "the Shedders" who . . . 'want to collect experiences, not possessions' (Quelch, 2002).

But what constitutes an experience? Wright has proposed a framework for analysing experience, which is made up of two parts. 'The first part is concerned with describing experience from four points of view which we refer to as the four threads. The second part is concerned with how we make sense in experience' (Wright, 2003). An experience is described in terms of its structure, our sensory and emotional engagement, and the actions and events over time and in a place. 'Experiences do not present themselves to us ready-made, people actively construct them through a process of sense making' (Wright, 2003). This process consists of our expectations, responses, interpretations and reflections of an experience.

In *The Experience Economy*, Joseph Pine and James Gilmore suggest that companies are moving beyond services, into experiences, in order to differentiate themselves. 'The experience economy is a new stage of economic offering.' Pine suggests that consumers must be drawn into the offering much like a viewer watching a theatre performance, but the viewer must also be an actor and participate: 'The consumer – sorry, the guest, must be drawn into the offering such that they feel a sensation. And to feel the sensation, the guest must actively participate' (Pine *et al.*, 1999).

Elements that stimulate our senses (sight, touch, sound, taste and smell) form our experiences of our environment. Products have always engendered some kind of sensorial quality. Intelligent materials will provide a new array of sensorial qualities, which will have an impact both on how we experience our surroundings and how we interact with them. An intelligent world will be one in which our interactions with products become ever more intuitive, using materials and systems that are responsive to our methods of communication, such as through touch and the use of body language. Intelligent materials will improve the control we have over our material environment and facilitate our creative interaction with it as we seek to be co-creators, tailoring experiences to correspond to our various moods. Gershenfeld has stated that:

> in the laboratory and in product development pipelines, information is moving out of traditional computers and into the world around us, a change that is much more significant than the arrival of multimedia or the Internet because it touches on so much more of human experience. (Gershenfeld, 1999)

As a result, we are seeing the rise in the industrial design community of what is termed experiential design, a method that is engaged with the value of the experience the user derives from using a product that will eventually become ubiquitous in all areas of design practice. To build a body of knowledge with which to frame experiences, designers have to investigate how people use, engage and feel about things and places.

11.4 Design and development: multidisciplinary collaboration

If the predictions are correct, a life where intelligence is embedded into everything is going to make the process of developing products and materials much more complex than traditionally has been the case, as it will encroach on different sectors and involve many complex issues. 'Dozens of smart fabrics and interactive textiles-enabling technologies are under development today, yet few of the OEMs or end-users of SFIT-enabled (smart fabrics and interactive textiles) solutions know about these technologies' (VDC, 2003). In other words, formal channels of communication do not currently exist between the users and the developers, nor between the discrete sectors that will be involved in developing materials. The development of electronic textiles is underway in defence research agencies in the USA, but it may be some time before such textiles enter the commercial domain.

To address this lack of dialogue, mechanisms are needed to bring the different industries closer. For example, the author coordinates the network Smart Textiles for Intelligent Products, funded by the Engineering and Physical Sciences Research Council, UK. This network is a think tank for future intelligent or smart consumer products and applications in the context of societies and markets of the future. It seeks to create new formal channels of communication by bringing together all sectors that will be involved in the design, development and production of the products into a new hybrid community. Those include: application-based industries (sports, clothing, medical, automotive, gaming, architectural and interior environments), defence agencies, cognitive science, social science, computing, electronics, electrochemistry, textile and fibre engineering, the design community (fashion, textile, industrial, interior designers and architects), economists, business and markets specialists, and lifestyle trend forecasters. Careful consideration has to be given to facilitating liaisons between people of different sectors, as each speaks a different language and thinks in different ways. Also, these various industries have discrete cultures, whose timescales for development and production vary hugely. Through its workshops, the smart textiles network aims to break down these barriers. There are two levels of workshop. The first comprises technology special-interest workshops, which bring together users with the different fields of expertise required to develop smart textile platforms, such as textile actuators and sensors and displays. This workshop brings together the potential users with the developers of materials. The second type of workshop explores the bigger picture by looking at projections of the future of society, lifestyle, work, travel, economics and markets, which members use to brainstorm collectively about possible product scenarios for the future. It is envisaged that these scenarios will set a trajectory for the development of smart textiles.

11.5 A new language for textiles: combining the real and the virtual

Given the vision of dematerialising information and communication technology (ICT), and the fact that much of our living environment is made from textiles that are familiar and 'friendly', soft and tactile, the ICT industries are now expressing a keen interest in textiles. One of the concerns consumers often have about new technology products is their tendency to become more sophisticated, thereby making them difficult to use and adapt to. Therefore, high tech must not make our products more complicated by having more components; rather, high tech should become seamlessly integrated into everyday objects without altering their character, and enhance their function. This section looks at how the design of textiles and clothing can converge with information and communication technology, and play a key role in this emerging genre of intelligent products and environments.

Embedded intelligence will completely alter the relationship that people have with everyday products and environments. No one yet knows how people will react to or engage with technology worn on their bodies or integrated into their homes. Intelligence has no tangible form. Seamlessly embedded intelligence will change the way designers design and develop products, as the focus will no longer be solely on the physical form of the product, but about intangible features, such as the notion of experience and emotional fulfilment that affect all the senses. If people are increasingly going to search for challenging experiences, then designers must create a context for experience. The designers' task will become one of giving form to virtual content. In the realm of human–computer interaction, content is seen as service, experience, communication and access to information. Intelligence will give the designer greater scope for creativity, a new tool with which to explore and apply computer intelligence in new ways. It will also challenge the role of the designer. Will the designer become more of a facilitator and enable users or wearers to be inventive by means of the technology? When the interface between the user and technology becomes truly ubiquitous, where the user does not need to understand the inner workings of the technology, then the potential of people to be inventive with the technology will be enhanced.

There are a number of fundamental issues that need to be understood before technology can be applied to the body or integrated into people's homes in a way that is truly meaningful. Addressing these involves a multidisciplinary effort, which examines conventions and cultures of product use and experience.

> As the human–computer interface becomes more pervasive and intimate, it will need to explicitly draw upon cognitive science as a basis for

understanding what people are capable of doing. User experience and situation should be integrated into the computer system design process. (Selker and Burleson, 2000)

In examining the established roles and places that objects have in our everyday lives, and the psychology of interaction with and cognition of products, the human sciences and the designer of clothing and products should be involved in the designing of systems and materials from the outset.

Referring again to Smart Paper, Gershenfield has stated that:

> Choosing between books and computers makes as much sense as choosing between breathing and eating. Books do a magnificent job of conveying static information; computers let information change. We're just learning how to use a lot of new technology to match the performance of the mature technology in books, transcending its inherent limits without sacrificing its best features. The bits and atoms belong together. The story of the book is not coming to an end; it's really just beginning. (Gershenfield, 1999)

By building on what we know of existing products, like the Smart Paper example, we can begin to extend and augment the utility of clothing and other textile products, such as furnishings and building materials. 'Augmented reality offers designers of electronic products the powerful notion that we can interact with electronics through everyday real objects, with their inherent richness of interaction' (Djajadiningrat *et al.*, 2000).

Researchers in textile, clothing and interior environment design at CSM, London, are exploring the interface of intelligent technology as well as extending the application of computer intelligence, building on existing cultures of clothing and products. Disseminating concepts and ideas in the public domain has the potential to excite market demand and, therefore, prompt the development of materials. CSM researchers are exploring the application of existing technologies, such as wireless communications, textile antennas, chromic display materials, textile switches, textile circuits and microcomponent welding technology, in new ways. For example, the author's research, funded by the Arts and Humanities Research Board, UK, is entitled 'Interactive and Experiential Design in Smart Textile Products and Applications'. It is well known that textiles have their own language that is at once tactile, sensorial and visual, which textile and fashion designers have traditionally exploited to engineer or express a look, a concept or idea, by carefully composing and manipulating the many facets of its special vocabulary. The language of textiles will be expanded exponentially as a result of the integration of electronic technologies to build smart textile systems. This research aims to discover what new codes of interaction and experience will arise when textiles are transformed from a passive into an active, intelligent state. The notion of clothing as a tool box will be conceptualised. These tool boxes will

Interaction design in smart textiles clothing and applications 231

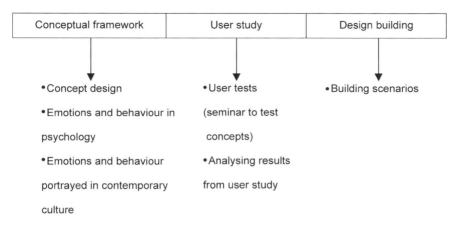

11.1 Design methodology.

enable the user to experience a sense of being creative, to communicate more expressive and emotional messages, and to engage in social interaction and gaming. The visual look and haptic qualities of smart clothing and interior environments can be customised by the user through non-verbal channels of communication that are intuitive, such as by gesturing and touching. The research seeks to rationalise the function of ICT-content together with conventional interaction with everyday clothing and their cultures, and map one onto the other to ascertain what new interactions and experiences will arise.

The interaction design process is an iterative one, where known conventions about how people use and experience products are used to map a conceptual framework for form of content and experience. The method used here makes reference to the methodology explored by the University of Art and Design, Helsinki, Finland (Mattelmäki and Keinonen, 2001). The framework is based on observations and research on how people use, interact with and experience conventional clothing and interior environments: the social psychology of people's clothing use and behaviour; sensory perception of textiles; how people communicate their emotions through non-verbal channels, such as body language; how people communicate through wireless communications systems, such as mobile phones and the internet; how expressive communication is portrayed in contemporary culture in terms of language, moods and colours. User group tests are based on a seminar in order to test conceptual assumptions. The results are fed back into the framework, which is then used to build prototype designs. These are also tested on user groups and are again fed back into the framework. This concept design method is now gaining momentum, and is illustrated in Fig. 11.1.

The following design scenarios suggest a way in which textiles and clothing can converge with information and communication technology. The scenarios make reference to a newly emerging area of expressive interaction, interpersonal and haptic (sense of touch) communication, and gaming. For example, the Super Cilia

Skin is an interactive membrane that was developed at the MIT Media Lab. The skin functions as a computer output device capable of visual and tactile expression, allowing gestures to be seen or an image to be felt via an array of actuators mounted onto an elastic membrane.

> Most computational tools rely on visual output devices. While such devices are invaluable, influential studies in neurophysiology have shown that physical experience creates especially strong neural pathways in the brain. When people participate in tactile/kinesthetic activity, the two hemispheres of the brain are simultaneously engaged . . . assuring that new information will be retained in long-term memory. (Raffle, 2003)

Another example of interpersonal communication fusing haptic technology is ComTouch, again developed at the MIT Media Lab. ComTouch is a handheld device that translates finger pressure into vibration, thereby augmenting voice channels of communication. 'A device that conveys touch might allow for more expressive interactions' (Chang et al., 2002).

11.5.1 Tools for remote interpersonal communication

The aims of this scenario are to develop: a clothing concept that facilitates the sending of expressive messages to friends or partners by conveying a sense and experience of touch or presence through clothing; and clothing that is a mobile aid that facilitates the expression of other aspects of human communication, supporting the user's need for subtle communication and complementing existing channels of communication.

This scenario is built on some conventions and cultures of clothing and textiles as expressive media. One of the main attributes of textiles is that their huge range of tactile qualities (cool/warm, hard/soft), as well as acoustic properties, has a certain effect on the way people feel and respond to them. Sensory science or psychophysics is an emerging area of experimental psychology that was first applied to product areas such as food. It is now being applied to textiles to measure people's subjective experiences of textiles when touched. Psychophysics is being developed for use in e-commerce applications where the tactile qualities of textiles need to be conveyed visually. Touch is an important part of human interaction and communication, for example, warmth and affection are often conveyed through touch. Also, people communicate through gesture or body movement, which constitutes a type of language, a 'language of emotions' according to Darwin (Darwin, 1965). Clothing is an emotional medium; it envelops us, is our second skin and is in some way an extension of our body. 'Dress is the way in which individuals learn to live in their bodies and feel at home in them. Dress is . . . an intimate experience of the body' (Entwistle, 2000). Therefore, this scenario builds on the close bond we have with our clothing, which connects people through touch and gesture and allows more expressive interactions to take place remotely. The

Table 11.1 Tools for remote interpersonal communication

Emotional expression or mood	Sending message/ switching action	Receiving message/change effected (in recipient's clothing)	
		Display	Actuator
Warmth	Gentle squeezing of arm	Yellow	Squeezing sensation
Love	Stroking of arm	Red	Soft tactile
Affection	Embrace/arms wrapped around the wearer	Yellow	Hugging sensation

conceptual framework for remote interpersonal communication is illustrated in Table 11.1.

Sending the message: The signal that is sent is based on the relationship that the sender has with the recipient. The message can be purely non-verbal, an expression of how the sender is feeling or a feeling that he or she wants to convey, or illustrative, reinforcing verbal messages over the telephone. The sender's interactions (or switching actions) with his or her clothing are based on conventions of touch and gesture associated with expressive communication and/or touch or gesture that serve to back up verbal communication. For example, emotional expressions that demonstrate affection could be conveyed through touch and gestures such as an embrace or stroking an arm/sleeve. The garment consists of pressure-sensing and gesture-sensing textiles connected to a textile antenna by a textile circuit, to which a communications chip is welded. When the pressure-sensing textile is pressed or stroked, or when the gesture sensor senses a gesture or movement, a signal containing a code is sent to the communications chip. Each type of pressure or gesture is assigned a code.

Receiving the message: The change effected in the recipient's clothing is based on translating the expressive meaning of the touch or gesture into colour or tactile configurations, realised in a chromic display material. The pressure or gesture code is picked up by the receiver's antenna, causing the communications chip to send a signal to the display on which a colour appears. The emotion expressed by the sender is translated into a colour, which is based on known colour psychology and cultural values of colour. Colours are purported to have emotional, physical and behavioural values; for example, in many Western cultures the emotional value of red is love, vitality, courage, passion and danger. Colours have positive and negative effects on us, caused by their energy entering our bodies. By being able to effect a change in colour in the receiver's clothing, the sender can either let the recipient know how he or she is feeling or influence his or her mood. The receiver's garment is composed of a textile antenna and communications chip, which are connected either to display materials or to actuator textiles by textile circuits.

Touch and gesture can be conveyed more literally using actuator textiles; for example, an embrace sent could be conveyed through the contraction of the fibres of the textile, causing the garment to hug the body. Touch can also be tactilely sensed, for example, where different haptic qualities of textiles can be conveyed. Configurations of haptic qualities are based on research conducted on the sensory properties of various types of textiles. The concept could also be extended to include interior environments, where a sense of presence can be conveyed via furniture and furnishings.

11.5.2 Tools for social interaction and social gaming

'Social interactions are the focus of our existence. We are social animals, and for any technology to be useful, it must eventually support socialization' (Ark, 1999). The aim of this scenario is to develop clothing concepts that facilitate social interaction by provoking and eliciting emotional responses. People can interfere and interact with the clothing of others in the vicinity by changing the visual appearance (colour, pattern), tactile quality or shape of the clothing.

This scenario is built on some conventions of clothing as an expressive medium and on the social nature of humans. Clothing facilitates social interaction, as it is a means of making the body social. It can create a sense of belonging or enable anonymity through a process of managing personal appearance to form a total composite image, thereby provoking responses from others. Clothing can be used as a channel of communication, where 'one person would *say* something to another person with the intention of effecting some change in that other person. . . . The effect on the receiver is important in that it is the effect on the receiver that constitutes social interaction' (Barnard, 1996). The effect can be an emotional response, change of behaviour or state of mind. People express and communicate their emotions through their behaviour and body language. 'Bodily, non-verbal communication operates within a social context, but also . . . the messages conveyed by bodily expression are about the society itself' (Douglas, 1971). The conceptual framework for this scenario is based on people's clothing behaviour, as well on explorations into the social behaviour of people. This scenario is also built on the emerging area of personal electronic data exchange and on the fact that gaming is becoming more of a social activity. The vCard is an electronic business card, which is a new means of sending business cards to people via electronic devices. Developers of electronic games, such as Sony, are looking into making gaming less of a solitary activity and more of a social one. The conceptual framework for social interaction and social gaming tools is illustrated in Table 11.2.

Active clothing can augment people's interactions with each other in social spaces by eliciting responses. The wearer can attract attention to himself or herself, let other people know how he or she feels or simply have fun by changing the appearance of someone else's clothing; for example, by sending someone a colour.

Table 11.2 Tools for social interaction and social gaming

Emotional expression or mood	Sending message/ switching action	Receiving message/change effected	
		Display	Actuator
Being playful	Squeezing or tapping arm, shoulder	In recipient's clothing: orange	In recipient's clothing: squeezing sensation
Sending personal message	Positive body language	In recipient's clothing: yellow	In recipient's clothing: pressing sensation
Social gaming	Gaming actions	In recipient's clothing: hit spots	In recipient's clothing: hit sensation

As in the previous scenario, interactions between people are based on gestures and actions as communicators of emotion, which serve to trigger changes in either the sender's or the recipient's clothing.

11.5.3 Tools for creativity and gaming

The aims of this scenario are to develop textile concepts of clothing and the interior environment that facilitate a sense of being creative by allowing the user to be a co-creator. The user or wearer customises the visual appearance (colour, pattern), tactile quality or shape of the textile, thus giving the wearer a sense of self-expression.

This scenario is built on some conventions and cultures of clothing. Clothing facilitates individualistic expression, allowing individuals to differentiate themselves and to declare their uniqueness.

> It seems that more and more people are becoming addicted to the feelings they get when they do wear something new. Those feelings may be of increased or reinforced uniqueness or of pleasure in presenting a different appearance to the world. Individuals may also derive aesthetic pleasure from either 'creating personal display' or from appreciating that of others. (Bernard, 1996)

Clothing can also serve to reflect, hide or generate mood. Sometimes, by expressing a mood, the wearer can influence other people's moods. This scenario is also built on the celebrity culture or limelight syndrome, prevalent in Western cultures, and on the field of gaming. Companies such as Sony and Microsoft are looking at ways of making gaming more interactive and fun; for example, researchers at Sony

Table 11.3 Tools for creativity and gaming

Emotional expression or mood	Switching action	Change effected	
		Display	Actuator
Customising aesthetics: personal display	Virtual paint-box: drawing hand down sleeve effects brushed colour changes	User's clothing: any colour	User's clothing: change in size, shape, tactile quality
Role play/fantasy: celebrity, icon, film character	Role play actions	Mediated environment: provides context	User's clothing: force feedback
Game/sport: e.g. Jujitsu	Jujitsu actions	Mediated environment: provides game context User's clothing: hit spots	User's clothing: force feedback

are investigating the use of gestures as game commands to replace the joystick. The conceptual framework for creativity and gaming tools is illustrated in Table 11.3.

Clothing can be customised to alter the aesthetics, colour, pattern, shape and size of clothing according to the size, taste and mood of the wearer. The wearer can be expressive by changing aesthetics and effects, for example, changing light, colour and patterns, shapes and textures. Gaming or role playing can be enhanced by mediated environments, where the space is also smart or active. Mediated spaces have no interface; people simply speak and act as they normally do, and the space understands them. In a mediated environment the user can live out fantasies and engage in games, sport, performance or role playing/acting in their own homes, for example, where body actions are read and sent to a wall display. By monitoring performance, the clothing can augment sports activities, for example, by embedding sensors in running shoes or in the flooring, facilitating feedback on technique. Clothing can also enhance performance by providing extra strength, which may be of particular benefit for those with injuries or disabilities.

11.6 Technology enablers

The term 'smart' is used to define a material that reacts in a particular way when exposed to stimuli such as environmental changes, for example, temperature or electrical currents. The following technologies are examples of developments and applications in the field to date, and illustrate the types of textiles and systems that will be used to explore the design scenarios. A range of conductive textiles has

Interaction design in smart textiles clothing and applications 237

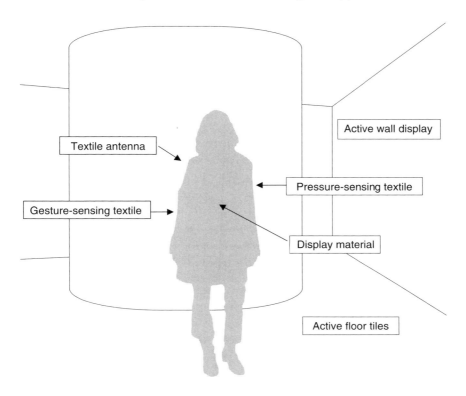

11.2 Location of active materials in clothing and in an environment.

been tested for their properties as antennas and circuits by Cliff Randell, Matthew Chalmers and Henk Müller who are members of the UK computing research project, Equator, in collaboration with the author. Figure 11.2 illustrates the location of active materials in the concepts of active clothing and environments.

11.6.1 Conductive textiles: switching and sensing

The first attempts to make smart textiles dealt largely with the area of integrating electronic functionality into textiles for the transmission of signals, using conductive fibres such as carbon with mounted electronic components to build smart textile systems. Fabrics made from conductive fibres are used to sense pressure for keyboards, such as the PDA keyboard by ElecSen, UK. This type of conductive textile can sense pressure as well as the degree and type of pressure, so that the system can differentiate between pressing and stroking. Softswitch, UK, is made from conductive textile materials and a quantum tunnelling composite (QTC) with unique pressure-controllable switching properties. Applications exist: in pressure sensing, for example car seats; in switching, for example, switches that allow electronics to function within clothing; in the medical field, for example, to monitor the position of hospital patients while in bed or in wheelchairs to prevent sores.

11.6.2 Conductive textiles: transmission of signals

Philips Research, UK, used knitted conductive fibres to produce a textile mechanical sensor that can measure gestures and movements. Infineon, Germany, has woven textile circuits from silver-coated polyester fibres.

11.6.3 Conductive textiles: generation of heat

Changes in the colour of fabrics can be triggered by sending an electrical impulse through a conductive fabric on which thermochromic ink is printed, as was done by International Fashion Machines, USA (a spin-off from MIT) in their development of Electric Plaid. Gorix, USA, has used conductive fabrics as heating elements for temperature-controlled car seats and performance outerwear.

11.6.4 Conductive textiles: antennas

Conductive copper and silver-coated nylon fibres have been shown to act as antennas. Philips Research, UK, and Randell, Baurley, Chalmers and Müller have tested such textiles.

11.6.5 Wireless networks: using the human body as a network

Researchers from NTT DoCoMo Multi-media Labs and NTT Microsystem Integration Labs in Japan have used the body's electrical field to transmit data at an Ethernet-like 10 megabits per second. The network, ElectrAura-Net, uses a combination of the electric field that emanates from the body and a similar field emanating from special floor tiles to transmit information.

11.6.6 Biometric sensing

The scenarios could be extended further by the use of textile systems that sense the physiological signals of the wearer (sweat sensors, respiration monitor, pulse/heart monitor, body temperature sensor, brain activity monitor), which would allow the textile system (clothing and environment) to recognise how the wearer is feeling and respond accordingly.

11.7 Future technology enablers

Many electroactive polymers are currently being developed by the electrochemistry industry, a convergence of electronics and chemistry. The electrochemistry sector is developing electroactive polymers for what are being termed plastic electronics. Examples of these include polymer light-emitting diodes (LEDs), which are printed by ink jet onto flexible substrates for flexible displays and packaging by companies such as Plastic Logic, UK, a spin-off from the Cavendish Laboratory, Cambridge University, UK. The Cavendish Laboratory is also developing plastic electronic circuits from polymer transistors, as well as sensors and memories for smart electronic devices. The future of truly smart textiles lies in the potential for technology convergence, where these electroactive polymers or molecular electronics are processed into or fabricated onto, fibres and fabrics. If transferred to the textiles industry, these polymers will make possible the production of soft intelligent textile products that will permit a broad spectrum of functions and capabilities.

11.7.1 Electroactive polymers: light-emitting

Visson has pioneered a light-emitting fabric, where conductive fibres are coated with electroluminescent polymers. The electrons within the polymer become excited upon the application of an electric charge, thus generating photons, which is coloured light. Clemson University, USA, has developed dynamic colour-responsive chameleon fibre systems. Hollow polyaniline, an inherently conductive fibre, is coated with an electrochromic substance; the hollow fibre membrane transports ionic charge carriers to the electrochromic coating, acting as dopants to turn the colour on or off.

11.7.2 Electronic ink (E-Ink)

The principal components of electronic ink are millions of tiny microcapsules. Each microcapsule contains positively charged white particles and negatively charged black particles suspended in a clear fluid. To form an E-Ink electronic display, the ink is printed onto a sheet of plastic film that is laminated to a layer of circuitry. The circuitry forms a pattern of pixels that can then be controlled by a display driver. E-ink is printed using existing screen-printing processes onto virtually any surface, including glass, plastic, fabric and even paper. Philips and E-Ink are to launch a new flexible electronic display in 2004.

11.7.3 Electroactive polymers: power

One of the biggest problems in wearable and integrated electronic technology is power. Power Paper's core technology is an innovative process that enables the printing of thin, flexible and environmentally friendly energy cells onto a polymer film substrate, by means of a simple mass-printing technology. Power Paper cells are composed of two non-toxic, widely available commodities, zinc and manganese dioxide, which can be printed onto virtually any substrate.

11.7.4 Electroactive polymers: actuators

Actuator polymers are a new breed of polymer that responds to external electrical stimulation by displaying a significant displacement in shape or size. When a current is applied, the polypyrrole polymer's accordian-shaped molecules stretch out like human muscles; when the current stops, the polymer contracts. Electroactive actuator polymers are currently being investigated by institutions such as the US military for the next generation of battle suits. The Jet Propulsion Laboratory, NASA, USA, is using them to attempt to develop artificial muscles for applications such as miniature gripper arms for robots to be deployed in explorations. If fibres could be fabricated from actuator polymers, textiles that change their shape and surface texture could be realised.

11.7.5 Future fabrication: nano

The application of intelligent functions into textiles using the aforementioned electroactive polymers will rely on nanotechnology. 'The currently existing multibillion-dollar world market influenced by nanotechnology is supposed to affect nearly any industry sector in the future' (Institute of Nanotechnology, 2003). Nanotechnology will be the next industrial revolution following the knowledge age. Future developments in biotechnology and nanotechnology could make it possible to alter the basic characteristics of almost any substance. 'From a commercial viewpoint, such developments would enable manufacturers to tailor their products and product performance characteristics exactly to the needs of their customers' (Textiles Intelligence, 2002). Nanotechnology is the creation of functional materials, devices and systems through the control of matter on the nanometre scale. Research is being conducted on modifying the surfaces of fibres, and on grafting materials onto fibres to create multifunctional, responsive and adaptive fibres. The main aim is to tailor a hybrid nanolayer of polymer film that will afford a number of functions and properties, for example, colour change. The electronics sector is developing nanowires that are grown from vapours of atomic ingredients and that act as diodes or other electronic components.

11.8 Conclusions

The development of fundamental textile technology platforms for application to a new genre of intelligent products and environments will require cross-sectoral consultation and collaboration. Also, to ascertain where the market pull that will influence the direction of the development of materials might come from, concepts and ideas should be floated in the public domain. The information and communication technology community will need to collaborate with those in design and the human sciences, if intelligent products and systems are to be embraced by consumers.

11.9 Acknowledgement

This chapter is a result of study being conducted by Sharon Baurley, supported by the Arts and Humanities Research Board through the AHRB's Fellowships in the Creative and Performing Arts scheme, UK.

Central Saint Martins College of Art & Design is the largest of the five colleges that constitute The University of the Arts London. The University of the Arts London is the largest institution for education and research in art and design in Europe.

11.10 References

Ark W S (1999), 'A look at human interaction with pervasive computing', *IBM Systems J.*, **38**(4), 504–508, New York, USA.
Barnard M (1996), *Fashion as Communication*, London, Routledge.
Chang A, O'OModhrain S, Jacob R, Gunther E and Ishii H (2002), 'ComTouch: design of a vibrotactile communication device', *Proceedings from Symposium on Designing Interactive Systems*, London, ACM Press, 213–320.
Darwin C (1965), *The Expression of the Emotions in Man and Animals*, London, Chicago, University of Chicago Press.
Djajadiningrat J P, Overbeeke C J and Wensveen S A G (2000), 'Augmenting fun and beauty: a pamphlet', *Proceedings from Designing Augmented Reality Environments*, Elsinore, Denmark.
Douglas M (1971), 'Do dogs laugh? A cross-cultural approach to body symbolism', *J. Psychosomatic Res.*, **15**.
Entwistle J (2000), *The Fashioned Body: Fashion, dress, and modern social theory*, Cambridge, Polity Press.
Gershenfield N (1999), *When Things Start to Think*, London, Hodder and Stoughton.
Institute of Nanotechnology, UK, www.nano.org.uk
Macdonald N (2003), *About: Interaction Design*, a Design Council paper series on design issues, London, UK.
Mann S (1998), 'Keynote address', *Proceedings from First International Conference on Wearable Computing*, Fairfax, VA, USA.
Marzano S and Arts E (2003), *The New Everyday, Views on Ambient Intelligence*, Rotterdam, 010 Publishers.
Maslow A (1970), *Motivation and Personality*, 2nd edition, New York, Harper & Row.
Mattelmäki T and Keinonen T (2001), 'Design for brawling – exploring emotional issues for concept design', *Proceedings from The International Conference on Affective Human Factors Design*, London, Asean Academic Press.
Picard R and Healey J (1997), 'Affective wearables', *Proceedings from the First International Symposium on Wearable Computers*, Cambridge, MA, USA.

Pine B J, Pine B J II and Gilmore J H (1999), *The Experience Economy*, Harvard, Harvard Business School Press.
Quelch J (2002) 'Too much stuff', The World in 2002, London, *The Economist* Newspaper Ltd.
Raffle H (2003), 'Super cilia skin: an interactive membrane', *Proceedings from Conference on Computer Human Interaction*, Ft. Lauderdale, USA.
Rekimoto J (2001), 'Gesture wrist and gesture pad: unobtrusive wearable interaction devices', Interaction Laboratory, *Proceedings from Fifth International Symposium on Wearable Computers*, Zurich, Switzerland.
Selker T and Burleson W (2000), 'Context-aware design and interaction in computer systems', *IBM Systems J.*, **39**(3/4).
Textiles Intelligence (2002), 'Future materials: technical textiles for the information age', *Technical Textiles Markets*, UK, **51**, 4th quarter.
VDC (2003), *Smart Fabrics and Interactive Textiles: a global market opportunity assessment*, Venture Development Corporation, USA.
Weiser M (1996), www.ubiq.com/hypertext/weiser/UbiHome.html
Wright P (2003), 'A framework for analysing user experience', *Usability News*, UK, 2 April 2003.

11.11 Sources of further information

Affective Computing, MIT, MA, USA, http://affect.media.mit.edu
Cavendish Laboratory, Cambridge University, UK, www.phy.cam.ac.uk
Central Saint Martins College of Art & Design, London, UK, www.csm.arts.ac.uk
Department of Kansei Engineering, Faculty of Textile Science and Technology, Shinshu University, Japan, www.tex.shinshu-u.ac.jp/faculties/kansei/kansei_e.html
E-Ink Corporation, Cambridge, MA, USA, www.eink.com
Eleksen Limited, Iver Heath, UK, www.eleksen.com
EPSRC Network, *Smart Textiles for Intelligent Products*, CSM, London, UK, www.smartextiles.co.uk
Gorix Ltd., Southport, UK, www.gorix.com
Hickson III M L, Stacks D W and Moore N-J, *Nonverbal Communication Studies and Applications*, Los Angeles, Roxbury
IBM Almaden Research Centre, San Jose, CA, USA, www.almaden.ibm.com
Infineon, Munich, Germany, www.infineon.com
Interaction Design Department, Royal College of Art, London, UK, www.interaction.rca.ac.uk
Interactive Institute, Ivrea, Italy, www.interaction-ivrea.it/en/index.asp
International Fashion Machines, Cambridge, MA, USA, www.ifmachines.com
Jet Propulsion Lab., NASA, Pasadena, CA, USA, ndeaa.jpl.nasa.gov
Kaiser S B (1997), *The Social Psychology of Clothing*, New York, Fairchild Publications
Massey P J (2000), 'Fabric antennas for mobile telephony integrated within clothing', *London Communications Symposium, Proceedings from the Annual London Conference on Communication* (UCL), London, http://www.ee.ucl.ac.uk/lcs/prog00.html
Media Lab., MIT, Boston, MA, USA, www.media.mit.edu
NTT *DoCoMo*, Multi-media Labs, Tokyo, Japan, www.nttdocomo.com
Philippe F, Schacher L and Adolphe D C (2003), 'The sensory panel applied to textile goods – a new marketing tool', *J. Fashion Marketing and Management*, **7**(3).
Plastic Logic, Cambridge, UK, www.plasticlogic.co.uk

Polhemus T (editor) (1978), *Social Aspects of the Human Body*, New York, Penguin.
Power Paper Ltd., Petah Tivka, Israel, www.powerpaper.com
Randell C, Baurley S, Chalmers M and Müller H (2004), 'Tools for wearable computing', *Proceedings from First International Forum on Applied Wearable Computing*, Bremen, Germany, 63–74.
School of Materials Science and Engineering, Clemson University, Clemson, SC, USA, mse.clemson.edu/index.htm
Sensatex smart shirt, New York, USA, www.sensatex.com
Smart Paper by Gyricon, Ann Arbor, MI, USA, www.gyriconmedia.com/SmartPaper.asp
Softswitch, Ilkley, UK, www.softswitch.co.uk
Sony Computer Science Laboratory, Tokyo, Japan, www.csl.sony.co.jp
The Disappearing Computer Initiative, European Union, www.disappearing-computer.net
Things That Think Group, MIT, Boston, MA, USA, ttt.media.mit.edu
Valdez P and Mehrabian A (1994), 'Effects of color on emotions', *J. Experimental Psychology*, **123**(4), 394–409.
vcard, Internet Mail Consortium, Santa Cruz, CA, USA, www.imc.org/pdi
Wearable Computing Project, MIT, MA, USA, www.media.mit.edu/wearables
Xerox Parc Research Centre, Palo Alto, CA, USA, www.parc.xerox.com

Index

active clothing 234
actuating fabrics 67–71, 181–2
   carbon nanotube fibre actuators 71
   conducting polymer fibre actuators 70–1
   dielectric elastomer wearable actuators 68–70
actuator polymers 70–1, 240
affective computing 226
American Sign Language 76
Anderson localisation 141
antennas 238
applications
   of communication apparel 159
   of electroceramic fibres 54–5
   for flexible displays 171–2
   in multimedia 75–6
   of smart textiles 124–33
   of wearable electronics 8–10, 101–2
arctic clothing 179
   Reima survival suit 216–18
audio interfaces 3

bending of optical fibres 111–15, 116–19
Berry numbers 17–18
biometric sensing 238
biphasic fibres 164–5
block-based technology 2
Bluetooth 5, 211–13
body movement detection 72–4
bonded structures 110
Bragg grating structure 124–5, 136–7, 139

cables and wiring 5, 119, 205, 217–18
cameras 1–2
carbon black 29

carbon fibres 83
carbon loaded rubber (CLR) sensors 63, 65–7
carbon nanotubes 29, 71
central processing units (CPUs) 183, 201
ceramic fibres
   applications in intelligent apparel 54–5
   epoxy 1-3 composites 45–9
   parallel and series model 49–54
   polymer 1-3 composites 49–54
   PT (lead titanate) 42–5
   PZT (lead zirconate titanate) 41, 42–5
chameleon fibres 164
chemoresistivity 61
CLCs (cholesteric liquid crystals) 144
clothing 9–10, 178–9
   active clothing 234
   colours 233, 238
   concept 232–4
   and creativity 235–6
   psychophysics 232–3
   and social interaction 234–5
   see also communication apparel; smart clothing
CLR (carbon loaded rubber) sensors 63, 65–7
coating of nanofibres 33–4
coherent backscattering 141
colour sources 136, 138
colours of clothing 233, 238
communication apparel 155–74
   applications 159
   automatic functioning 158
   connection problems 162–3
   data processing 162
   design 160
   displays 161–2, 163–73
   electronic parts 160

energy sources 162, 163
  interfaces 160–2
  manual functioning 158
  peripherals 160–2
  sensors 161
communication technologies 4–5, 8
  infrared communication 212–13
  radio frequency (RF) systems 213–14
  remote interpersonal tools 232–4
  requirements in smart clothing 202–3, 204
computing systems
  affective computing 226
  interaction design 226
  pervasive computing 225
  ubiquitous computing 224–5
  wearable computing 177–8, 198, 214–16, 225
  see also electronic textiles
concept model 199–200
conducting polymer fibre actuators 70–1
conductive carbon fibres 83
conductive fabrics and fibres 7–8, 13–14, 81, 164, 205–7, 237–8
  electromechanical properties 81, 82–4
  shielding 205–6
  see also smart clothing; smart textiles
conductive polymer coatings 83
connections 162–3, 205, 206–7
copper wires see wiring and cables
cotton fabrics 63–7
CPU (central processing unit) 183, 201
creativity 235–6
crosstalk 194
cultural implications 10–11, 226

DARPA 177
data management technologies 5–6
data processing 162
data transfer 202–14
  cables 205, 217–18
  communication model for smart clothing 202–3
  connectors 205, 206–7
  electrically conductive fibres 205–7
  external 208–9
  internal 210–14
  personal space 210–14

  requirements 203
  wired 203–7
  wireless 207–10, 238
design processes 160, 231–2
development of technology 1, 223–40
diameter of fibres 14–15, 16–19
dielectric constant 34–7
dielectric elastomer wearable actuators 68–70
display matrix design 166–71
display technologies 4
  in communication apparel 161–2, 163–73
  fibre-harvesting ambient light-reflective 138–40
  flat panels 150
  flexible displays 151, 164–72
  optical fibre (OFFD) 165–71
  reflective displays 139–40
  side-emitting fabrics 151
  textile-based 151, 164–5
dye molecules 143

E-broidery project 181
E-Ink 239
economic implications 10
electric wiring and cables 5, 119, 205, 217–18
electroactive fabrics 59–79
  actuating fabrics 67–71, 181–2
  for health care 71
  as kinaesthetic interfaces 76–9
  for motion capture 71–6
  multimedia applications 75–6
  sensing fabrics 62–7
  see also smart textiles
electroactive nanofibres 21–34
electroactive polymers 182, 239–40
electroceramic fibres 41–55
  applications in intelligent apparel 54–5
  ceramic fibre/epoxy 1-3 composites 45–9
  ceramic fibre/polymer 1-3 composites 49–54
  parallel and series model 49–54
  PT (lead titanate) fibres 42–5
  PZT (lead zirconate titanate) fibres 41, 42–5

246   Index

electrodeless coatings 33–4, 84
electrokinetic polymers 61
electroluminescent fibres and fabrics 145–51
electromagnetic interference (EMI) shielding 82, 180
electromechanical properties 81–102
  of ceramic fibre/epoxy 1-3 composites 49
  of conductive textiles 81, 82–4
  of PPy-coated conductive fibres/yarns 84–99
  of stainless steel fabric 99–101
electromechano-chemical polymers 62
electronic ink 239
electronic polymers 4
electronic textiles 179–84
  actuators 181–2
  networks 180–1, 184–94
  power supply 183
  processors 183
  sensors 181–2
  signal transmission 186–94
electrospinning 15–21, 38
  Berry numbers 17–18
  diameter of fibres 14–15, 16–19
  rotary electrospinning 19–20
  of self-assembled yarn 20–1
  stretching process 15–16
  yarn and fabric formation 19–21
electrostatically generated nanofibres 13–38
  coating of nanofibres 33–4
  electroactive nanofibres 21–34
  electrospinning 15–21
  nanocomposites 29–33
  pyrolysis of nanofibres 33–4
  ultra-low dielectric constant nanocomposite film 34–7
electrostatics 62
electrostriction 62
embedded intelligence 228, 229
embedded wiring 5
EMI shielding 82, 180
energy management technologies 6–7
energy sources 6–7, 162, 163, 183, 202, 239–40
epoxy 1-3 composites 45–9

ESBG (electrically switchable Bragg grating) 139–40
Experience Economy 227
external data transfer 208–9

fabric-based displays 163–73
fabric-based interfaces 3
fabric-based sensors 3
fashion applications 9–10
FBG sensor 124–5
fibre-based technology 2, 106, 107
  see also optical fibres
fibre-harvesting ambient light-reflective displays 138–40
flat panel displays 150
flexible displays 171–2
  applications 171–2
  optical fibre flexible displays (OFFDs) 165–71
  textile-based 151, 164–5
  see also display technologies
flexible substrates 150
FOLEDs (flexible organic light emitting devices) 145–51
France Telecom 182
frequency characterisation 192–3
fuel cells 6

gaming 235–6
Gershenfield, N. 227, 230
Gilmore, James 227
Gore-Seam® shielding 206
graphite nanoplatelets 29
GSM systems 4–5, 208–9

healthcare 8–9, 11–12, 71
  monitoring shirts 124
  smart fabrics for 71
  telesurgery 76
heat generation 238
heating jacket 218–19
hierarchy of needs 226–7
HIPERLAN 210
home applications 9–10
HPDLC (holographic polymer dispersed liquid crystal) screens 4, 139–40
human–computer interaction design 226

Index

ICD+ 2
ICPs (intrinsically conductive polymers) 14, 21–9, 37
impedance measurement 188–90
impedance simulation 190–2
implementation model 200–2
implications of wearable technology 10–12
industrial applications 10
Industrial Clothing Design Plus 2
infrared communication 212–13
inks 84, 239
input interfaces 3
integrated circuits 7–8
integrating machinery 110–11
integrating optical fibres 119–24
intelligent apparel *see* smart clothing
interaction design 226
interfaces 2–4, 160–2
  kinaesthetic interfaces 76–9
  in smart clothing 201–2
interference effects 138, 141
internal data transfer 210–14
interpersonal communication tools 232–4
iridescent films and fibres 138
ISM bands 210–11

joysticks 76

K materials 34–5, 38
kinaesthetic interfaces 76–9
kinetic energy 6
knitted structures 99–101, 107–8, 122
knitting machines 115
knowledge age 224–6

Lawandy model 143–4
layers of clothing 199–200
LCD (liquid crystal display) screens 4
lead titanate (PT) fibres 42–5
lead zirconate titanate (PZT) fibres 41, 42–5
LEDs (light emitting diodes) 169, 182, 239
leisure applications 9–10
Lifeshirt 179
light emitting diodes (LEDs) 169, 182, 239

light generation 136
long-range communications 4–5
Lycra fabrics 63–7

machinery for integrating optical fibres 110–11
McKibben effect 69
magnetic storage systems 5–6
magnetoresistivity 61
magnetostriction 61
Mann, Steven 225
manufacturing smart textiles 115–24
Maslow's hierarchy of needs 226–7
material properties 187
medical applications *see* healthcare
metal fibres 83
metallic coated fibres 83
metallic nanoparticles 29
military applications 10
modified parallel and series model 49–54
Morpho butterflies 138
motion capture 71–6
  body movement detection 72–4
MP3 players 2
MPD (modal power distribution) sensors 125–9, 131
Multi-Fibre Arrangement 223
multimedia applications 75–6

nanocomposites 29–33
nanofibres
  coating 33–4
  electroactive 21–34
  electrospinning 15–21
  manufacturing methods 15
  pyrolysis 33–4
  ultra-low dielectric constant nanocomposite film 34–7
nanotechnology 240
networks 180–1, 184–94
  electrical characterisation 186–94
  geometry of textiles 184–6
Newton–Raphson iteration 112
non-woven structures 110, 116

OLEDs (organic light emitting devices) 145–51

Olofsson model 114
optical fibres 106, 107–9, 116–19, 165, 180
  bending of fibres 111–15, 116–19
  display matrix design 166–71
  fabric displays 163–73
  FBG sensor 124–5
  flexible displays (OFFD) 165–71
  integrating 110–11, 119–24
  side-emitting displays 151
  for transmitting light 139
  weaving 166
optical storage systems 6
opto-amplification 140–5
organic light emitting devices (OLEDs) 145–51
orthogonal interlace calculations 112–15
output interfaces 3–4

PAN (personal area networks) 5
parachutes 124–33
parallel and series model 49–54
PBGs 136–8, 144–5
Peirce geometrical model 112, 114
peripherals in communication apparel 160–2
personal position manager (PPM) 19–20
personal space communication 202–3
personal space data transfer 210–14
pervasive computing 225
Philips jacket 179
photoconductivity 61
photoelectricity 61
photonic band-gap materials (PBGs) 136–8, 144–5
photonics 136–52
  electroluminescent fibres and fabrics 145–51
  fibre-harvesting ambient light-reflective displays 138–40
  opto-amplification 140–5
  textile-based flexible displays 151
Picard, Roz 226
piezoelectric materials 6–7, 41
piezoelectricity polymers 61, 62
piezoresistivity 61
Pine, Joseph 227
plain weave yarns 108–9

PLEDs (polymer light-emitting diodes) 4
polyaniline fibres 16, 21–9
polymers
  actuator polymers 70–1, 240
  ceramic fibre/polymer 1-3 composites 49–54
  conductive coatings 83
  electroactive 182, 239–40
  electromechano-chemical 62
  ICPs (intrinsically conductive polymers) 14, 21–9, 37
  PLEDs (polymer light-emitting diodes) 4
  for sensing applications 60, 61–2
  see also PPy
position manager (PPM) 19–20
power supplies 6–7, 162, 163, 183, 202, 239–40
PPM (personal position manager) 19–20
PPy sensors 62, 63–5
PPy-coated conductive fibres/yarns 62, 63–5, 81–2, 84–99
  performance under strain 93–5
  performance under tension 87–93, 95–9
processors 183, 201
psychophysics 232–3
PT (lead titanate) fibres 42–5
pyroelectricity 61
pyrolysis of nanofibres 33–4
PZT (lead zirconate titanate) fibres 41, 42–5

radio frequency (RF) systems 213–14
reflective displays 139–40
Reima survival suit 216–18
remote interpersonal communication tools 232–4
rotary electrospinning 19–20

satin weave yarns 109
self-assembled yarn 20–1
semi-conductive textiles 81
sensing fabrics 62–7
sensors 2–3, 181–2, 237–8
  biometric sensing 238
  body movement detection 72–4
  CLR (carbon loaded rubber) 63, 65–7

Index   249

and communication apparel 161
FBG sensor 124–5
fibre optic sensors 106, 107
MPD (modal power distribution) 125–9, 131
polymers for sensor design 60, 61–2
PPy sensors 62, 63–5
sensing fabrics 62–7
SOFTswitch 182
temperature sensors 218–19
WearNET 179
see also smart textiles
shielding conductive fibres 82, 180, 205–6
short message service (SMS) 208
short-range communications 5
side-emitting displays 151
sign language 76
signal integrity 194
signal transformations 181–2
signal transmission 186–94, 238
  crosstalk 194
  frequency characterisation 192–3
  impedance measurement 188–90
  impedance simulation 190–2
  material properties 187
  signal integrity 194
  transmission line configuration 187–8
silicon chips 7
silicone rubbers 68
single-wall carbon nanotubes (SWNTs) 29–33
smart clothing 9–10, 158–9, 198–202
  central processing unit (CPU) 201
  communication requirements 202–3, 204
  concept model 199–200
  electroceramic fibres 54–5
  health monitoring shirts 124
  heating jacket 218–19
  implementation model 200–2
  layers 199–200
  power management 202
  survival suits 179, 216–18
  uniforms 124
  user interfaces 201–2
  see also clothing; communication apparel; data transfer

smart parachutes 124–33
smart textiles 105–6, 107–11
  applications 124–33
  bending of optical fibres 111–15
  for healthcare 71
  integrating optical fibres 119–24
    machinery used 110–11
  intelligent apparel 158–9
  as kinaesthetic interfaces 76–9
  knitted structures 99–101, 107–8, 122
  manufacturing 115–24
  for motion capture 71–6
  non-woven structures 110, 116
  woven structures 109, 122
Smart Textiles Network 228–9
SMS (short message service) 208
social implications 10–11, 234–5
SOFTswitch 182
sol-gel process 41, 45
solid-state storage systems 6
stainless steel fabric 99–101
storage technologies 5–6
stretching process 15–16
survival suit 179, 216–18
SWNTs (single-wall carbon nanotubes) 29–33

technological development 1, 223–40
telecommunications 171
telesurgery 76
temperature sensors 218–19
template method 15
textile geometry 184–6
textile networks *see* networks
textile-based flexible displays 151, 164–5
textiles industry 223
thermoelectricity 61
thermoresistivity 61
transformation of signals 181–2
transmission line configuration 187–8
twill weave yarns 109

ubiquitous computing 224–5
ultra-low dielectric constant nanocomposite film 34–7
underwear layer 199–200
uniforms 124
user interfaces *see* interfaces

vibration interfaces 3
video games 76
visual interfaces 4
voice synthesis 3–4
VSSP (viscous suspension spinning process) 41

weak localisation 141
wearable computing 177–8, 198, 214–16, 225
wearable electronics 8–10, 101–2
Wearable Motherboard 182
WearNET 179
weaving machines 115–16
weaving optical fibres 166
web formation structures 110

Weiser, Mark 224
wired data transfer 203–7
wireless data transfer 207–10, 238
wiring and cables 5, 119, 205, 217–18
WLAN (wireless local area networks) 209–10
woven structures 109, 122
WPAN (wireless personal area networks) 211–12
wrist devices 1–2

Xerogel film 34–5

yarn and fabric formation 19–21
Yo-Yo 201